Advances in Intelligent Systems and Computing

Volume 266

Series editor

Janusz Kacprzyk, Warsaw, Poland

For further volumes:

http://www.springer.com/series/11156

G. P. Biswas · Sushanta Mukhopadhyay
Editors

Recent Advances in Information Technology

RAIT-2014 Proceedings

Springer

Editors
G. P. Biswas
Sushanta Mukhopadhyay
Computer Science and Engineering
Indian School of Mines
Dhanbad
Jharkhand
India

ISSN 2194-5357   ISSN 2194-5365 (electronic)
ISBN 978-81-322-1855-5   ISBN 978-81-322-1856-2 (eBook)
DOI 10.1007/978-81-322-1856-2
Springer New Delhi Heidelberg New York Dordrecht London

Library of Congress Control Number: 2014933389

Printed on acid-free paper

Springer is part of Springer Science+Business Media (www.springer.com)

Preface

Information (and Communication) Technology (IT), which emerged at the end of nineteenth century as one of the most potent growth areas, is an outcome due to a fundamental change in our society from Industry to Information/Knowledge-based activities. IT has ever-growing impact upon the society and it continues to lead, shape, and support all affairs of the human beings. It is indeed widely recognized that the societal development depends on the successful progress of IT. Although it emerged as a convergence of computer and communication technologies, in broad sense it refers to all aspects of computing as well as the technological needs in academia, business, government, public-service sectors, and other kinds of organizations for automation and administration. It thus encompasses a vast majority of diverse fields in science, technology, management, and many more. The present proceedings, which is edited based on the selected innovative original research articles of the *2nd International Conference on Recent Advances on Information Technology (RAIT-2014)*, organized by the Department of Computer Science and Engineering, Indian School of Mines, Dhanbad, India, not only enhances the IT advancement, but also helps the IT professionals all over the world for their new creativities/research in IT.

We arranged for publishing the conference proceeding by Springer as a book series *Advances in Intelligent Systems and Computing* for maximum possible reachability to the readers worldwide. It has four parts comprising—(1) *Network and Security*: It contains four chapters, two of which deal with designing of routing protocols for mobile ad-hoc and wireless sensor networks, and other two are on network application and security issues; (2) *Image Processing Applications*: This part comprises of four contributory chapters on various applications of image processing namely, granulometric image analysis based on first digital geometric approach for image-based quality monitoring, document image processing for character and hand-written script recognition, and DCT-based image processing for cryptographic application; (3) *Topics in Algorithm Design*: Three innovative chapters mainly designing efficient algorithms on shortest path spanning tree of a given graph, dynamic checkpoint strategy usable in computational grid and computation of DFT in multi mesh network; and (4) *Mathematical Computing*: This part has five chapters that cover the following areas—precise estimation of population mean in nonresponsive condition, modeling of epidemic spread under Allee effect, rainfall prediction based on pattern similarity and K-NN technique, dimensionality reduction of gene expression for designing rule-based classifier and Eigen value estimation based on Radial Basis Function.

Finally, editors look forward that the edited volume would earn accolade from the researchers working in the related areas.

Organizing Committee

Chief Patron	Sri P. K. Lahiri, IAS (Retd.), Chairman, GC and EB, ISM, Dhanbad
Patron	Prof. D. C. Panigrahi, Director, ISM, Dhanbad
General Chair	Dr. C. Kumar/HOD (CSE), Prof. G. P. Biswas
Program Chair	Prof. G. P. Biswas
Program Co-Chair	Dr. S. Mukhopadhyay
Organizing Secretary	Dr. C. Kumar
Treasurer	Dr. SachinTripathi

Advisory Committee

Prof. Chin-Chen Chang, Feng Chia University, China
Prof. Rajkumar Buyya, University of Melbourne, Australia
Prof. Demetrios Sampson, University of Piraeus, Greece
Dr. Narayan C. Debnath, Winona State University, USA
Prof. Subir Bandyopadhyay, University of Windsor, Canada
Prof. Sajal K. Das, University of Texas at Arlington, USA
Prof. Shi-Jinn Horng, National Taiwan University of Science and Technology, Taiwan
Prof. K. K. Raymond Choo, University of South Australia, Australia
Prof. Raj Jain, Washington University in St. Louis, USA
Prof. Goutam Chakraborty, Iwate Prefectural University, Japan
Dr. Neils Volkmann, Stanford-Burnham Medical Research Institute, CA, USA
Dr. Pradeep Atrey, University of Winnipeg, Canada
Dr. Fagen Li, School of Computer Science and Engineering, China
Dr. Subash C. Mishra, Scientist, Harvard University, USA
Dr. Sumita Mishra, Rochester Institute of Technology, NY
Prof. Bharat Bhargava, Purdue University, USA
Prof. Somenath Biswas, IIT Kanpur, India

Organizing Committee

Prof. A. K. Nirala, ISM, Dhanbad
Prof. Shalivahan, ISM, Dhanbad
Prof. G. Udaybhanu, ISM, Dhanbad
Prof. S. Bhattacharya (ECE), ISM, Dhanbad
Prof. G. P. Biswas, ISM, Dhanbad
Prof. Prasanta K. Jana, ISM, Dhanbad
Dr. C. Kumar, ISM, Dhanbad
Dr. S. Mukhopadhyay, ISM, Dhanbad
Dr. A. R. Dixit, ISM, Dhanbad
Dr. Somnath Chattopadhyay, ISM, Dhanbad
Dr. Rima Chatterjee, ISM, Dhanbad
Dr. Keka Ojha, ISM, Dhanbad
Dr. Rajani Singh, ISM, Dhanbad
Dr. Saumya Singh, ISM, Dhanbad
Dr. A. K. Behura, ISM, Dhanbad
Dr. Biswajit Paul, ISM, Dhanbad
Dr. Kaylan Chatterjee, ISM, Dhanbad
Dr. Bobby K. Antony, ISM, Dhanbad
Dr. Haider Banka, ISM, Dhanbad
Dr. Hari Om, ISM, Dhanbad
Dr. M. K. Singh, ISM, Dhanbad
Dr. Sukomal Pal, ISM, Dhanbad
Dr. Sachin Tripathi, ISM, Dhanbad
Dr. Arup Kumar Pal, ISM, Dhanbad
Mr. A. C. S. Rao, ISM, Dhanbad
Mr. A. Tarachand, ISM, Dhanbad
Mr. Dipankar Ray, ISM, Dhanbad
Mr. Rajendra Pamula, ISM, Dhanbad
Mr. D. Ramesh, ISM, Dhanbad
Mrs. S. V. Maheshkar, ISM, Dhanbad
Ms. Shweta R. Malwe, ISM, Dhanbad
Ms. Tanusree Kaibartta, ISM, Dhanbad
Dr. Pramod Mathur, ISM, Dhanbad
Mr. Rajesh Kumar Mishra, ISM, Dhanbad
Mr. Rakesh Soni, ISM, Dhanbad
Mr. S. Mitra, ISM, Dhanbad

Program Committee

Prof. V. K. Srivastava, ISM, Dhanbad
Prof. N. R. Mandre, ISM, Dhanbad
Prof. A. K. Pathak, ISM, Dhanbad
Prof. T. Sharma, ISM, Dhanbad

Prof. A. Sarkar, ISM, Dhanbad
Prof. T. K. Chartterjee, ISM, Dhanbad
Prof. N. M. Mishra, ISM, Dhanbad
Prof. C. Bhar, ISM, Dhanbad
Prof. J. K. Patttanayak, ISM, Dhanbad
Prof. K. Dasgupta, ISM, Dhanbad
Prof. A. K. Pal, ISM, Dhanbad
Prof. S. K. Maity, ISM, Dhanbad
Prof. V. Priye, ISM, Dhanbad
Prof. A. K. Mukhopadhyay, ISM, Dhanbad
Prof. P. K. Sadhu, ISM, Dhanbad
Prof. P. Pathak, ISM, Dhanbad
Prof. Prasanta K. Jana, ISM, Dhanbad
Prof. Shalivahan, ISM, Dhanbad
Prof. S. Gupta, ISM, Dhanbad
Prof. G. N. Singh, ISM, Dhanbad
Prof. R. K. Updhayay, ISM, Dhanbad
Prof. Debjani Mitra, ISM, Dhanbad
Prof. Somnath Pan, ISM, Dhanbad
Prof. G. P. Biswas, ISM, Dhanbad
Prof. A. S. Venkatesh, ISM, Dhanbad
Prof. P. S. Mukherjee, ISM, Dhanbad
Prof. Suhas Deb, CIT, Ranchi, India
Dr. Arnab Bhattacharya, IIT Kanpur, India
Dr. Debashis Ghosh, IIT Roorkee, India
Dr. Abdullah Bin Gani, University of Malaya, Malaysia
Dr. Basabi Chakraborty, Iwate Prefectural University, Japan
Dr. Rajiv Mishra, IIT Patna, India
Dr. Ashok K. Das, Center for Security, Institute of Information Technology, Hyderabad
Dr. A. K. Tripathi, IIT BHU, Varanasi, India
Dr. Anil K. Tiwari, IIT Jodhpur, Rajasthan, India
Dr. Debdeep Mukhopadhyay, IIT Kharagpur, India
Dr. Jatin K. Deka, IIT Guwahati, Assam, India
Dr. Sudip Mishra, IIT Kharagpur, India
Dr. Koliya Pulasinghe, Srilanka Institute of Information Technology, Srilanka
Dr. Somanath Tripathi, IIT Patna, India
Dr. G. Rama Murthy, IIIT, Hyderabad, India
Dr. A Kar, ISM, Dhanbad
Dr. Sandeep Mondal, ISM, Dhanbad
Dr. S. Das, ISM, Dhanbad
Dr. Dheeraj Kumar, ISM, Dhanbad
Dr. Ajay Mondal, ISM, Dhanbad
Dr. A. R. Dixit, ISM, Dhanbad
Dr. Chiranjeev Kumar, ISM, Dhanbad

Dr. Sushanta Mukhopadhyay, ISM, Dhanbad
Dr. Haider Banka, ISM, Dhanbad
Dr. Hari Om, ISM, Dhanbad
Dr. M. P. Singh, NIT, Patna
Dr. M. K. Singh, ISM, Dhanbad
Dr. S. P. Tiwari, ISM, Dhanbad
Dr. SachinTripathi, ISM, Dhanbad
Dr. Sukomal Pal, ISM, Dhanbad
Dr. Arup K. Pal, ISM, Dhanbad
Mr. A. C. S. Rao, ISM, Dhanbad
Mr. A. Tarachand, ISM, Dhanbad
Mr. Rajendra Pamula, ISM, Dhanbad
Mr. D. Ramesh, ISM, Dhanbad
Mrs. S. V. Maheshkar, ISM, Dhanbad
Ms. Shweta R. Malwe, ISM, Dhanbad
Ms. Tanusree Kaibartta, ISM, Dhanbad

Acknowledgments

It gives us immense pleasure to express our sincere gratitude to our chief patrons—Shri P. K. Lahiri, IAS (Retd.), Honorable Chairman, Executive Board and General Council, ISM and Prof. D. C. Panigrahi, Honorable Director, ISM, for their strong support and invaluable guidance to organize RAIT-2014. We are also genuinely indebted to all the members of our Advisory and Organizing Committees for their kind cooperation at various stages.

Our special thanks and acknowledgments to the author participants for attending and presenting their contributory works in the conference. We also express our deepest gratitude to the reviewers for taking so much pain in reviewing the articles. Our heartfelt thanks to all our invited speakers who have devoted precious time in spite of their extremely tight schedules.

A large number of student participants who attended our conference deserve thanks for their active, dynamic, and enthusiastic support without which the conference would not have been possible. The faculty members, officers, and other staff members of the department and Institute are also acknowledged for all kinds of support and cooperation.

Finally, our special thanks to ISEA (Information Security Education and Awareness) Project, funded by the Ministry of Communication and Information Technology (MoCIT), Government of India, for supporting us to execute our event.

Contents

Part I Technical Session-I: Network and Security

**Dynamic Clustering Based Hybrid Routing Protocol for Mobile
Ad Hoc Networks** 3
Shweta R. Malwe and G. P. Biswas

Design of Hybrid MAC Protocol for Wireless Sensor Network 11
Sankar Mukherjee and G. P. Biswas

Enabling Smartphone as Gateway to Wireless Sensor Network 19
Sourav Kumar Dhar, Suman Sankar Bhunia, Sarbani Roy
and Nandini Mukherjee

**Design of ECC-Based ElGamal Encryption Scheme
Using CL-PKC** 27
Manoj K. Mishra, S. K. Hafizul Islam and G. P. Biswas

Part II Technical Session-II: Image Processing Applications

**A Fast Digital-Geometric Approach for Granulometric
Image Analysis** 37
Sahadev Bera, Arindam Biswas and Bhargab B. Bhattacharya

**Word-Level Handwritten Script Identification
from Multi-script Documents** 49
Mallikarjun Hangarge and K. C. Santosh

**Selection of Graph-Based Features for Character Recognition
Using Similarity-Based Feature Dependency
and Rough Set Theory** 57
Sunanda Das, Suvra jyoti Choudhury, Asit Kumar Das and Jaya Sil

A Partial Image Cryptosystem Based on Discrete Cosine
Transform and Arnold Transform   65
Kshiramani Naik and Arup Kumar Pal

Part III   Technical Session-III: Topics in Algorithm Design

Swap Edges of Shortest Path Tree in Parallel   77
Anjeneya Swami Kare and Sanjeev Saxena

A New Approach for Parallel Discrete Fourier Transform
in Multi Mesh Network   87
Amit Datta and Mallika De

Dynamic Checkpoint Data Replication Strategy
in Computational Grid   95
Ramesh Babu and Subba Rao

Part IV   Technical Session-IV: Mathematical Computing

Estimation of Population Mean in Presence of Non Response
in Two-Occasion Successive Sampling   109
Arnab Bandyopadhyay and Garib Nath Singh

Modeling the Complex Dynamics of Epidemic Spread
Under Allee Effect   117
Parimita Roy and Ranjit Kumar Upadhyay

Rainfall Prediction Using k-NN Based Similarity Measure   125
Arpita Sharma and Mahua Bose

Dimension Reduction of Gene Expression Data for Designing
Optimized Rule Base Classifier   133
Amit Paul, Jaya Sil and Chitrangada Das Mukhopadhyay

Comparative Study of Radial Basis Function Neural Network
with Estimation of Eigenvalue in Image Using MATLAB   141
Abhisek Paul, Paritosh Bhattacharya and Santi Prasad Maity

Author Index   147

About the Editors

Dr. G. P. Biswas Professor, Department of Computer Science and Engineering, Indian School of Mines, Dhanbad, India. He completed his Ph.D. in Computer Science and Engineering from Indian Institute Technology, Kharagpur, with thrust research area on Testable Design of VLSI circuits. He did his B.Sc. Engineering and M.Sc. Engineering in Electrical and Electronic and Computer Science and Engineering, respectively. He has more than 25 years of teaching and research experience and published 90 research articles in Journals, Conferences and Seminar Proceedings. Apart from teaching, he has guided several B.Tech (CSE) and M.Tech (CA/CSE) projects, and executed external funded R&D projects. His main research interests include cryptography, computer networks with security, VLSI designs, and additive cellular automata.

Dr. Sushanta Mukhopadhyay Associate Professor, had joined the Department of Computer Science and Engineering, Indian School of Mines, Dhanbad in 2007 as an Assistant Professor. He did B.Sc. with Honours in Physics from Presidency College, Calcutta, B.Tech and subsequently, M.Tech in Radiophysics and Electronics from the University of Calcutta, India. He did Ph.D. from the Indian Statistical Institute, Calcutta. During 2001–2003 and 2004–2007, he worked at the Sanford Burnham Medical Research Institute, La Jolla, CA, USA and the Nanyang Technological University Singapore, respectively. His research areas and interests include image and video processing, fMRI, image/video compression, encryption, and watermarking. He has published a few articles in International Journals and guided M.Tech and Ph.D. students.

About the Department

The Department of Computer science and Engineering, set up in 1997, conducts academic programs leading to the award of B.Tech, M.Tech, and dual degrees in CSE, and Ph.D. in Computer Science and Engineering. The students of the department achieve high profession skills and are well placed in different IT Industries. The department has dedicated faculty members having expertise in diverse areas, some of whom are active members of reputed National and International Societies. The faculty members are actively involved in research and regularly publish research articles in reputed Journals and Conference Proceedings. They also execute different R&D projects funded by external agencies. The major research areas include Cryptography, Network Security, Computer Networks, Parallel and Distributed Computing, Internet Technology, Wireless and Mobile Communication, Software Engineering, Image Processing, Video-on-Demand, Soft Computing, Information Processing and Retrieval.

Visit http://www.ismdhanbad.ac.in/conference/cse/rait/index.php for more details.

Part I
Technical Session-I: Network and Security

Dynamic Clustering Based Hybrid Routing Protocol for Mobile Ad Hoc Networks

Shweta R. Malwe and G. P. Biswas

Abstract Routing in mobile ad hoc network must be efficient and resource-saving. Mobility causes frequent changes in the topology, which leads to inappropriate routing. One way to reduce traffic is to divide the network into clusters. Also, there should be a well-defined interconnection technique that guarantees optimized and stable network connectivity. Routing based on clustering provides an efficient source routing. In this paper, we propose an improved dynamic clustering based hybrid routing protocol which uses distributed spanning tree scheme for clustering that models the network and perform hybrid routing within the network. The proposed algorithms for clustering and routing have been compared with the existing scheme and results in the routing overhead reduction with optimized modelling of the network.

Keywords Mobile ad hoc network · Distributed spanning tree · Dynamic clustering · Hybrid routing protocol · Source routing

1 Introduction

A mobile ad hoc network (MANET) is a collection of wireless mobile nodes that dynamically forms a self-organizing network of mobile nodes. Such type of network faces challenges in terms of scalability, mobility, bandwidth limitations, and power constraints [1]. For MANET, there are three major categories for routing namely proactive (table-driven), reactive (on-demand) and hybrid (combined form

S. R. Malwe (✉) · G. P. Biswas
Dept of Computer Science and Engineering, Indian School of Mines, Dhanbad, India
e-mail: shweta.malve26@gmail.com

G. P. Biswas
e-mail: gpbiswas@gmail.com

G. P. Biswas and S. Mukhopadhyay (eds.), *Recent Advances in Information Technology*,
Advances in Intelligent Systems and Computing 266, DOI: 10.1007/978-81-322-1856-2_1,
© Springer India 2014

of proactive and reactive) routing. The efficiency of routing mainly depends on the organization of the underlying network which can be achieved by clustering within the network which divides the network into a set of clusters either dis-joint or overlapping that saves energy and communication bandwidth in the ad-hoc networks. The clustered topology provides three basic benefits. First, a cluster structure facilitates the spatial reuse of resources to increase the system capacity. Secondly, in the routing process, the set of cluster heads forms the backbone in routing process. Also, the network is organized systematically into a set of clusters that makes an ad hoc network appear more stable [2]. In this paper we discuss a new technique of clustering using distributed spanning approach which organizes the network into a set of disjoint hierarchical clusters. Further this clustering technique will be used for an efficient hybrid routing within the network were routing within the cluster will be proactive while between the clusters will be reactive in nature.

The rest of the paper is organized as follows. Section 2 contains a brief explanation of the literature work related to clustering and routing in MANET. Section 3 of the paper is organized describes the proposed Cluster based Hybrid Routing Protocol (CBHRP) discussing the major phases of Dynamic Clustering and Hybrid routing with explained algorithm and demonstration. The simulation of the proposed protocol with experimental results is described in Sect. 4. Finally, Sect. 5 concludes the paper.

2 Related Work

Clustering in ad hoc networks provide a systematic organization of an infrastructure less network. Various clustering techniques are proposed like identifier based clustering, connectivity based clustering, mobility aware clustering, Low cost of maintenance clustering, etc. [2]. Various routing protocol for mobile ad hoc networks have been proposed in the literature like AODV, DSDV, DSR, ZRP, etc. Cluster based routing is another approach of routing in MANET. Jiang, Li and Tay proposed Cluster Based Routing Protocol (CBRP) which is used to organize network into disjoint or overlapping clusters and route the data packets efficiently in the network with the help of gateways [3]. Although the network was cluster-organized but the main demerits of CBRP occurs during routing. Due to the support for unidirectional links, there is a large routing overhead in the network. Also, the overlapping clusters increase the routing control overhead. A Distributed Weighted Cluster Based Routing Protocol was proposed [4] where cluster head selection is done according to neighbourhood table and calculates the weight of node using weight calculation algorithm. Routing procedure is same as in CBRP. In [5], Krishna et al. proposed a cluster based approach for routing in dynamic networks by forming overlapping clusters using boundary nodes. Although the network gets cluster-connected using the specified algorithm, it lacks the efficiency parameter related to the number of clusters and routing overhead. Also, for organizing the network, Distributed Spanning Tree (DST) approach is proposed in the literature [6, 7], where network is organized as a spanning tree using probing procedures by the selected nodes of the network. Being a reactive protocol, CBRP has the advantage of minimizing the sending of message by limited flooding using clustering and is better approach for large ad hoc networks. But has demerits in some issues like size of the cluster, overlapping nature of clusters and overhead due to unidirectional links within or between clusters. Also, the instability of clusters leads to the frequent re-organization which directly affects routing and network performance. In the next section, we will be eliminating the above discussed demerits of the existing cluster based routing by proposing a new technique of clustering that help in an efficient routing in the network.

3 Proposed Cluster Based Hybrid Routing Protocol

The proposed Cluster Based Hybrid Routing Protocol (CBHRP) first organizes the network into a set of disjoint hierarchical clusters by Dynamic Clustering method using Distributed Spanning Tree (DST) scheme and then perform hybrid routing for packet delivery from source to destination within the network. CBHRP has two phases, cluster formation using Dynamic Clustering and Route Discovery.

(a) **Dynamic Clustering.** The proposed dynamic clustering method provides a set of disjoint hierarchical clusters in the network and is based on distributed spanning tree [6] scheme that helps to maintain the network organization and improve its performance during routing. The procedure of cluster formation will start with assumptions like, each node is assigned a unique identification number, *id*. Each cluster is required to have a minimum number of members. The two data structures are used for maintaining the information of nodes in the network. *Head Node Table (HNT)*: maintained by each head node and contains the information of other head nodes within its transmission range like *id* and distance of other head nodes, *Leaf Node Table (LNT)*: maintained by both head and leaf nodes and stores the information of the neighbouring leaf nodes (of the same cluster) in terms of *id* and distance. Leaf nodes also store head node information in *LNT*. After the cluster formation, if there exists a cluster having lesser number of leaf nodes than required, then the cluster head merges to the nearest cluster head thus forming a sub-cluster and includes a hierarchy within the new cluster as shown in Fig. 1a and b. Such hierarchy distributes the load between cluster heads within the single cluster and minimizes the overhead of packet forwarding.

The procedure of dynamic clustering consists of the following four procedures.

1. *Head Node Identification*: To initiate dynamic clustering, each node calculates the resource status according to the number of resources (like power, etc.) left. The nodes with higher resource status than threshold consider themselves as Head nodes. These head nodes will create *HNT*s and will latter becomes the cluster head of their respective clusters.

Fig. 1 **a** Cluster A and
Cluster B formed in a
network. **b** Cluster B merges
to Cluster A, forming
hierarchy within Cluster A

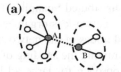 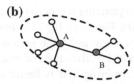

2. *Head Node Probing*: Each head node initiates probing by calling *HN_probe* procedure, to find the leaf nodes within its range of transmission by broadcasting a HELLO packet with a minimum transmission power. After probing, head nodes wait for a specific threshold time to get replied from leaf nodes.

3. *Leaf Node Reply*: The non-head nodes within range of a head node will reply the head node whose request came earliest, with a reply message by calling *LN_reply* procedure. It creates *LNT* to maintain information of the replied head node. If a reply message is accepted, the head node creates *LNT*, including the details of the leaf node.

4. *Cluster Merging, Creation and Maintenance*: Each cluster is required to have a minimum cluster size, which is determined by the number of leaf nodes in a cluster. If any head fails to have it, will call *merge* procedure to merge with the nearby cluster head, updates its *LNT* and form the hierarchy. Also a new cluster is created if a node is uninitialized for a long time or if the head node of a cluster is not connected to any of the other head nodes, by calling the procedure *create*. For maintaining the clusters, *HN_leave* and *LN_leave* procedures are called by head and leaf nodes, respectively.

(b) **Hybrid Routing.** Once the network is modelled as DST based hierarchical structures, the routing procedure is initiated by the source and following steps are proposed to discover the route from the source to its destination.

1. The source initiates the route discovery process by generating RREQ packet containing the destination id and enters the source id in the active route list.

2. If the source node is a leaf node then it forwards RREQ to the respective head node and enters the head node id in the active route list, else the source node is itself the head node and proceeds to the next step.

3. If the present head node is not the destination, then a procedure is called by the head node to check whether the destination is one of its leaf nodes.

4. If the current head node finds the destination in its Leaf Node Table (*LNT*), then RREQ is forwarded to the destination which in response generates RREP packet containing active route list, is transmitted to its source with the reverse route of the active route list and proceed to Step 7 for the next event.

5. If the current head node is unable to find destination in its *LNT*, it then broadcasts received RREQ packet to the head nodes that are within the range as mentioned in its Head Node Table (*HNT*).

6. The head nodes receiving the broadcasted RREQ packet will execute Step 3 to check for the destination.

7. As RREP packet is received by the source node, the path is established according to the active route list received in RREP packet from the destination and the data packets will be transmitted accordingly. Figure 2a and b shows the path of RREQ and RREP packet between a sender and its receiver throughout the network.

Local Route Shortening: The Leaf Node Table of the leaf nodes also stores the information of their neighbouring leaf nodes (of the same cluster). It will increase the efficiency of the routing process, specifically the intra-cluster routing. If the leaf nodes of the same cluster are within the transmission range, then direct communication is possible, instead of forwarding the data to the head node.

Figure 2a and b shows the path of RREQ and RREP packet between a sender and its receiver throughout the network.

Proposed algorithms for dynamic clustering procedure-

(A) Procedure *HN_initialize()*	**(D)** Procedure *LN_reply()*
/* N: Number of nodes, N[i]: selected nodes	/* msg is the reply message generated by leaf
HN[]: List of head nodes */	node */
Begin	Begin
Calculate resource parameter for each node	Set msg.id=N[i].id
If resource > threshold then Set N[i] = node	Set msg.pid=N[i].pid
Set HNT[]=N[i]	Send(msg,N[i].pid)
End	End
(B) Procedure *HN_probe(pmsg)*	**(E)** Procedure *HN_leave()*
/*pmsg is probe message generated by head node*/	/* N[i] is the leaf node of existing head node N[id]*/
Begin	Begin
Set pmsg.id=N[i].id	LN.delete(N[i])
Send(pmsg)// broadcast to non-head nodes	Set msg.id = N[i]
End	Set msg.pid = N[id]
(C) Procedure *create()*	Set msg.pr = 0
/* N1 is the victim head node or the uninitialized or	Set msg.leave = 1
unvisited node*/	Send (msg, N[i])
Begin	End
If N1 is the victim head node then	**(F)** Procedure *merge()*
Select a leaf node with high resource parameter	/* N1 is the head node to be merged with N2 head
from its LNT.	node*/
Exchange the status between head and leaf node.	Begin
End	Find the nearest head node from HNT.
If N1 was uninitialized then	Include the details of N1 in N2's LNT.
Consider N1 as head node and form its HNT.	Set LNT[N2].status='head'
Call HN_probe procedure.	Include the details of N2 in N1's LNT,
End	Set LNT[N2].status='head'
	End

4 Experimental Results

In this section, we presented the simulation results related to the performance of dynamic clustering and its effect on routing. Compared to the existing clustering for dynamic networks [5], the proposed one have a decrement in the number of clusters with increasing degree in the network with the total number of nodes,

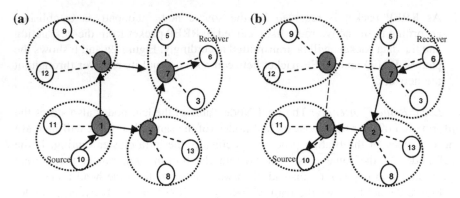

Fig. 2 **a** RREQ path from node 10 to node 6. **b** RREP path via reverse route

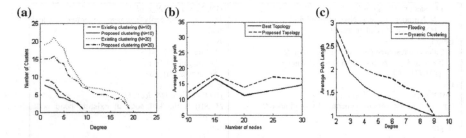

Fig. 3 Simulation results of dynamic clustering and hybrid routing

$N = 10$ and 20, as shown in Fig. 3a. The topology formed after the network organization using dynamic clustering is compared to the best topology formed on the same network. Although there is a stable network organization, the average cost per path increases in the network. Figure 3b illustrates the variation in the average cost per path with the increasing number of nodes in the network.

Figure 3c demonstrates the variation in the average path length with increasing degree in the network of 10 nodes. Path estimation in the non-clustered network is done using flooding which determines the shortest path between the two nodes participating in the route estimation. Average path length is computed as the average of the path lengths between each source and destination in the network. The proposed clustered network has a higher average path length when compared to flooding. The routing overhead is the ratio of the path length between a source and a destination estimated by the proposed cluster based routing and the best routing using flooding algorithm, respectively. It is observed that routing overhead is lesser than 1.75 which is not high and lesser than the value evaluated by the existing clustering method [5].

5 Conclusion

We have presented a novel Cluster Based Hybrid Routing Protocol (CBHRP) comprising the features of hierarchical clustering and hybrid nature of routing. Dynamic Clustering has been proposed, forms the first phase of the proposed routing protocol which models the network into a set of disjoint hierarchical clusters using Distributed Spanning Tree (DST). Once the network is organized into clusters, the routing procedure can be employed by Route Discovery procedure forming the second phase of the protocol. The experimental results show that the proposed clustering is optimized with respect to the mobility parameter of the nodes and forms lesser number of clusters than the existing clustering method. The average path length and cost for the route estimation is higher than flooding process but the routing overhead is decreased by 12.5 % which shows an improvement during routing.

References

1. Hoebeke, J., Moerman, I., Dhoedt, B., Demeester, P.: An overview of mobile ad hoc networks: Applications and challenges. J. Commun. Netw. **3**, 60–66 (2004)
2. Yu, J.Y., Chong, P.H.J.: A survey of clustering schemes for mobile ad hoc networks. EEE Commun. Surv. Tutorials **7**(1), 32–48 (2005)
3. Jiang, M., Li, J., Tay, Y.C.: Cluster based routing protocol (CBRP) functional specification. Internet draft, draft-ietf-manet-cbrp-spec-00.txt August (1998)
4. Chauhan, N., Awasthi, L.K., Chand, N., Katiyar, V., Chugh, A.: A distributed weighted cluster based routing protocol for MANETs. Wirel. Sens. Netw. **3**, 54–60 (2011). doi:10.4236/wsn. 2011.32006 Published Online February 2011
5. Krishna, P., Vaidya, N.H., Chatterjee, M., Pradhan, D.K.: A cluster-based approach for routing in dynamic networks. ACM SIGCOMM Comput. Commun. **27**(2), 49–64 (1997)
6. Paul, P.V., Vengattaraman, T., Dhavachelvan, P., Baskaran, R.: Modeling of mobile ad hoc networks using distributed spanning tree approach. Int. J. Eng. Sci. Technol. **2**(6), 2241–2247 (2010)
7. Paul, P.V., Vengattaraman, T., Dhavachelvan, P., Baskaran, R.: Improved data cache scheme using distributed spanning tree in mobile ad hoc network. Int. J. Comput. Sci. Commun. **1**(2), 329–332 (2010)

Design of Hybrid MAC Protocol for Wireless Sensor Network

Sankar Mukherjee and G. P. Biswas

Abstract Time division multiple access (TDMA) and frequency division multiple access (FDMA) are two popular multiple access technique that used in several wireless sensor networks. Both TDMA and FDMA have their own advantages and disadvantages. In this article, we have tried to use these two together for reducing the interference of inter-cluster and intra-cluster communication. We have considered a model of sensor network where, there are some high power nodes, which are placed in a regular hexagonal pattern and some ordinary sensor nodes which are placed randomly. High power nodes will act as clusterhead. Inside the cluster sensor nodes communicate with the clusterhead. TDMA and FDMA are being used for both time slot assignment and interference reduction. The proposed system is simulated to show the throughput of the system for different numbers of nodes. It is also shown how the packet loss of the whole system varies with low load and peak load. Our proposed system also compared with the contention based system and shown the improvement over the contention based system.

Keywords TDMA · FDMA · Frame · Cluster · MAC · Polling

1 Introduction

A sensor network typically consists of very large number of low cost sensor nodes which collaborates among each other to enable a wide range of applications. Unlike traditional data networks, communication protocol design in sensor

S. Mukherjee (✉)
Durgapur Institute of Advanced Technology and Management, Durgapur,
West Bengal, India
e-mail: sankar_mukherjee2000@yahoo.co.in

G. P. Biswas
Indian School of Mines, Dhanbad, India
e-mail: gpbiswas@gmail.com

G. P. Biswas and S. Mukhopadhyay (eds.), *Recent Advances in Information Technology*,   11
Advances in Intelligent Systems and Computing 266, DOI: 10.1007/978-81-322-1856-2_2,
© Springer India 2014

network is influenced greatly by their limited energy supply [1]. Therefore it is crucial for the sensor network protocols to be energy efficient in order to extend network lifetime. Traditional wireless medium access control (MAC) protocols such as IEEE 802.11 are designed for optimizing throughput, latency, and fairness without specifically concentrating on their energy usage. The asynchronous nature of these protocols prevents energy saving by not allowing wireless nodes to selectively put their network interfaces into low energy sleep modes [6]. Few works [2, 4, 6] has been done on reducing idle listening by powering off network interfaces when possible. A notable example of periodic active-sleep design is S-MAC [6]; in which node synchronize in active sleep cycle. A major drawback of SMAC is that it cannot have very small duty cycle and guaranteed bounded delivery latency because the basic medium access mechanism is contention based.

TDMA MAC protocols have built-in active-sleep duty cycle that can be leveraged for limiting idle listening, thus have better energy efficiency [3]. Here we have used TDMA and FDMA both as the MAC protocol. As we have used both TDMA and FDMA, that is why it is called hybrid MAC protocol.

2 System Model

The total area of the network is divided into some hexagonal cells. Some high power wireless nodes are deployed in this regular hexagonal pattern and placed in the three vertices of each hexagon as shown in Fig. 1. Sensor nodes are placed randomly in this area and Clusters are formed using the high power nodes. From the Fig. 1 it is seen that high power nodes are acting as clusterheads. The clusterhead nodes are generally used for data sensing just like normal sensor nodes as well as data aggregation. It also takes the responsibility to send the data to the sink node. So the clusterheads are used also for routing the data to the sink nodes. Clusterhead nodes have more energy than the normal sensor nodes.

Every sensor node must be under one cluster. Here some of the sensor nodes are belonging to more than one cluster boundary. So to resolve this problem, maximum received power among different neighboring clusterheads can be chosen as the clusterhead. The transmission range of each clusterhead node is considered as the length of the side of the hexagon. Transmission range of the sensor nodes is also same as the high power nodes. So the sensor nodes in each cluster are one hop away from its own clusterhead. Here each sensor node communicates to its own clusterhead only. There are two types of communication happen here, one is between sensor nodes to clusterhead and another is between clusterhead to clusterhead. So to overcome the interference for simultaneous communication among these nodes we have proposed TDMA and FDMA combinedly. In each cluster TDMA is used to assign the slots to the sensor nodes for communicating with their clusterhead. And for the communication between the clusterhead to clusterhead node, FDMA is used. Here every clusterhead nodes are being given fixed frequency channel for their communication and these frequencies are assigned in

Fig. 1 Network Structure of
clusterheads and sensor nodes

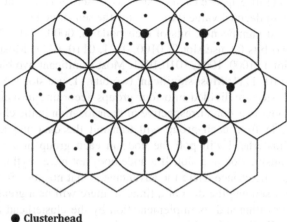

● **Clusterhead**

. **Sensor node**

such a way that there will not be any interference between them. These frequency channels are reused throughout the whole network in such a manner that the requirement of the frequency channels should be minimized. So the clusterheads and the sensor nodes can communicate independently without interfering each other.

Here we have considered that the network is using the ISM band. The whole bandwidth (83.5 MHz)) is divided into two parts in the ratio of 1:3. Higher part of the bandwidth is allocated to the clusters. According to the division higher part bandwidth is 62.62 MHz. This bandwidth is divided among the clusters in such a way that no two neighboring clusters use same frequency band. Sensor nodes use this frequency band along with TDMA technique to communicate with their clusterhead. The lower part of the bandwidth is used for communication between clusterhead to clusterhead. The lower part bandwidth is 20.87 MHz. Here the clusterheads use FDMA technique for communication between them.

2.1 TDMA and Time Slot Assignments

Here we have proposed that each cluster uses four time slots and each slot of length of one packet time. These four slots make one frame. Sensor nodes choose one slot among these four to send the data to the clusterhead. Slot assignment to the sensor nodes is done in a dynamic manner. Here we have used the polling method, asking to the sensor node for sending data in the slot. We have assumed that each cluster has sensor nodes which are 2^n in numbers. So out of this numbers of sensor nodes, ¼ th of the numbers use the first slot, ¼ th use the second slot and so on. Now which sensor nodes will use which slot that will be identified by their node id. Suppose each cluster has 16 sensor nodes and the node id of the nodes are from decimal value 0 to 15. Now the sensor nodes will use the slot according to the most significant 2 bits of their node id. 0000, 0001, 0010, 0011: Most significant two bits 00, use slot 0; 0100, 0101, 0110, 0111: Most significant two bits 01, use slot 1; 1000, 1001, 1010,1011: Most significant two bits 10, use slot 2; 1100, 1101, 1110, 1111: Most significant two bits 11, use slot 3. So all the sensor nodes in the cluster are divided into four groups according to their most significant bits. This division is done irrespective of their location in the cluster. In each frame only one node from each group will transmit the data and it is controlled dynamically. Clusterhead will poll the node in each group in a round robin fashion for data transmission in a slot. If a node does not have anything to send the next node will get the chance. Here the advantage is that no slot will be free and as every body is not sending the data at a time, so there will be a great energy saving. The polling algorithm and its implementation by the clusterhead are given below.

Polling Algorithm

1. Each clusterhead use four queues, Q1, Q2, Q3, and Q4 for four slots.
2. Insert node ids, eligible for sending in slot 1, in ascending order in the queue Q1.
3. Insert node ids, eligible for sending in slot 2, in ascending order in the queue Q2.
4. Insert node ids, eligible for sending in slot 3, in ascending order in the queue Q3.
5. Insert node ids, eligible for sending in slot 4, in ascending order in the queue Q4.
6. For slot No 1, pop the node ID from the Q1 and poll it for sending data in that slot. Insert node ID again into the Q1.
7. If node has data to send it will send the data,
8. else go to level 6 for next node.
9. The whole procedure is same for slot2, slot3 and slot 4 also. Slot 2 will be started after slot1; slot 3 will be started after slot 2 and so on.

In each cluster, clusterhead accumulates data that is send by the sensor nodes. It is the responsibility of the clusterheads to send the aggregated data to the sink node. Here we have assumed only one sink node. So to reach the sink node, clusterheads use their neighbor clusterheads. But communication between the clusterheads may create interference with each other, so they use FDMA as MAC protocol in the link layer. On the other hand when sensor nodes of different clusters communicate with their clusterhead, then intercluster interference may occur. So each cluster is assigned different frequency band to overcome the intercluster interference.

2.2 Frequency Assignment

Here we have assigned different frequency to different clusterheads. In the Fig. 2 the connection pattern of the clusterhead nodes are shown. Frequencies are assigned to the clusterhead nodes such that there will not be any interference between them.

Theorem 1: *Minimum 9 frequency bands are required to assign to the cluster-heads so that there will not be any interference.*

Proof: From Fig. 2 it is seen that the shaded area is a rectangle and this pattern repeats the whole figure. If we see the shaded rectangle the centre node is surrounded by four smaller rectangles. In fact every node is surrounded by maximum four smaller rectangles. And total numbers of nodes in all the four rectangles is eight. These eight nodes and the centre node should use different frequency band. So minimum number of frequency bands required for the clusterheads is 9(f1–f9).

Each clusterhead node is assigned a frequency channel such that its one hop and two hop neighbor does not use the same frequency channel. These frequency channels are basically use the lower part of the whole bandwidth. So the lower part of the bandwidth (20.87 MHz) is divided into equal 9 channels and each channel bandwidth is 2.31 MHz. These 9 channels are reused in the whole network.

On the other hand each cluster is also given different frequency band. Here each cluster can be thought of a cell of cellular network. Like cellular network here also the frequency reuse factor is 7.

Theorem 2: *Seven frequency band is sufficient to assign to the clusters so that there will not be any interference of nodes belonging to different clusters.*

Proof: Here from Fig. 1 it is seen that each cluster is surrounded by six clusters that means there are at most 6 neighboring clusters of each cluster. When the sensor nodes of one cluster communicate with its clusterhead, it may interfere with the communication of other sensor nodes which belong to neighbor clusters. So if the six neighboring clusters and the centre cluster use different frequency band, there will not be any interference in inter-cluster nodes. That is why minimum seven frequency band is sufficient to assign the clusters so that there will not be any interference of nodes belonging to different clusters.

To implement it, the higher part of the bandwidth (62.62 MHz) again divided into seven parts and each cluster will be given one part of it and the assignment is done in such a way that all six neighboring clusters and the cluster itself will get different frequency band each of 8.94 MHz. Inside each cluster, sensor nodes communicate with the clusterhead. So there will be interference among the sensor nodes if more than one sensor transmits at the same time. To overcome intra cluster interference problem here we have proposed TDMA in the MAC layer. So in the higher part of the bandwidth we have used FDMA and TDMA combindly.

Fig. 2 Connection pattern of
the clusterheads and
Frequency assignment

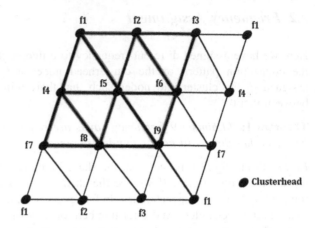

3 Simulation Result

The node mobility and traffic generations are simulated using Omnet++ 4.0 discrete event system simulator [5]. The nodes are distributed over an area of 600×600 m^2. The numbers of nodes has been taken 16 and 32 in each cluster. Range of the transmission of the high power nodes as well as sensor nodes are considered 100 units. Simulation has been done on finite random sensor networks using the proposed FDMA-TDMA scheme. In the simulation we have considered that the network is using the 2.45 GHz ISM band and data size of the sensor packet is 1 Kb.

Simulation is done on our proposed model and the contention based model. In the contention based model nodes will send data as they want and it may interfere with other neighboring nodes. So to send data, nodes have to contend with other nodes. In Fig. 3a we have shown how data transmission varies with time. It is found that our proposed model has more throughput than the contention based system. With time as the packet waiting time is more for contention based model that is why our proposed model gives better throughput than the contention based model.

In the second experiment we have seen how packet drop varies with total packet send. In Figs. 3b and c we find that our proposed model works better than the contention based model. Here we have drawn the curve for packet loss with successful packet send in the system in different simulation time with fixed number of nodes. Figure 3b has been drawn when number of nodes in each cluster is 16 and Fig. 3c has been drawn when number of nodes in each cluster is 32. Packet loss in the case of contention based model is more. Every packet has a TTL value. If a packet has to wait more time than TTL, the packet will be expired. That is why in the contention based model the packet drop rate is more than or proposed model. As we increase number of nodes in the whole system, contention will be more for contention based model which causes more packet drop. Similarly in the case of our proposed model as we increase number of nodes packet waiting time will be

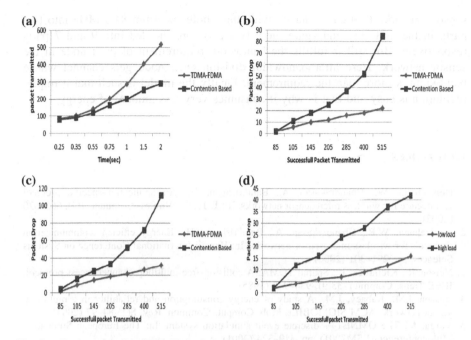

Fig. 3 **a** Packet transmitted with Time. **b** Packet drop with successful packet transmission for 16 nodes in each cluster. **c** Packet drop with successful packet transmission for 32 nodes in each cluster. **d** Packet drop while low and high load

increased because number of slot is fixed. Which causes more packet drop due to packet's TTL value is expired. In the Fig. 3c, when the number of nodes per cluster has been increased to 32, packet drop also increases compare to the Fig. 3b where number of nodes is 16 per cluster.

Finally we have observed how the packet drop varies with low load and high load in our system. Here load means packets generated in the system. We have assumed that packet generated in the system follows Poisson distribution formula. In our simulation, every sensor node generates 5 packets/sec in low load and 10 packets/sec in high load. From Fig. 3d it has been seen that when load is low packet drop is low compared to high load in the system. When load is high, packets has to wait more than when load is low. That is why high load create more packet drop than low load.

4 Conclusions

The proposed technique is to use frequency division and time division in a cluster based sensor network to reduce the interference in the inter cluster and intra cluster communication. It achieves less energy consumption which is most important for sensor network. Though we have divide the whole spectrum 84.3 MHz into two parts in the first step and again these two parts are divided into 9 and 7 parts respectively. But still it fulfills the bandwidth requirement of each node in the sensor network. Here after bandwidth division each node gets channel whose bandwidth is more than the requirement. Finally the model is such that it implementation is easy and that is why it consumes very less amount of energy.

References

1. Heinzelman, W., Chandrakasan, A., Balakrishnan, H.: An application-specific protocol architecture for wireless microsensor networks. IEEE Trans. Wireless Commun. 1(4), 660–670 (2002)
2. Heinzelman, W.R., Chandrakasan, A., Balakrishnan, H.: Energy efficient communication protocols for wireless microsensor networks. In: Hawaii International Conference on Systems Sciences, pp. 291–301 (2000)
3. Nelson, R., Kleinrock, L.: Spatial TDMA: A collision-free multihop channel access protocol. IEEE Trans. Commun. 33(9), 934–944 (1985)
4. Reason, J.M., Rabaey, J.M.: A study of energy consumption and reliability in a multi-hop sensor network. ACM SIGMOBILE Mob. Comput. Commun. Rev. 8(1) 84–97 (2004)
5. Varga, A.: The OMNET++ discrete event simulation system. In: The European Simulation Multiconference (ESM2001), pp. 319–324 (2001)
6. Ye, W., Heidemann, J., Estrin, D.:An energy-efficient mac protocol for wireless sensor networks. In: INFOCOM 2002, vol. 3, pp. 1567–1576 (2002)

Enabling Smartphone as Gateway to Wireless Sensor Network

Sourav Kumar Dhar, Suman Sankar Bhunia, Sarbani Roy and Nandini Mukherjee

Abstract The communication between Wireless Sensor Nodes and the Internet would require a gateway as these two network works on different standards. Mobile phone may be utilized to bridge this gap. It may act as a gateway for Wireless Sensor Nodes to connect with Internet and vice versa. Also this approach allows the smartphone to enrich the data collected from the Wireless Sensor Nodes before sending it to the Internet. In this paper, a low-cost solution is presented with open-source software and hardware that allows a wide range of smartphones to connect with sensors in a pervasive manner.

Keywords WSN · IoT · Cloud · Arduino · Smartphone

1 Introduction

In the era of Internet of Things (IoT) [1], every device is supposed to be connected with each other. Generally, Wireless Sensor Networks [2] are designed in such a way that the access to the network is through a central server or workstation which dispatches the data to the end users. Maintaining such design implementations

S. K. Dhar (✉) · S. S. Bhunia
School of Mobile Computing and Communication, Jadavpur University, Kolkata, India
e-mail: souravkrdhar@gmail.com

S. S. Bhunia
e-mail: bhunia.suman@gmail.com

S. Roy · N. Mukherjee
Department of Computer Science and Engineering, Jadavpur University, Kolkata, India
e-mail: sarbani.roy@cse.jdvu.ac.in

N. Mukherjee
e-mail: nmukherjee@cse.jdvu.ac.in

G. P. Biswas and S. Mukhopadhyay (eds.), *Recent Advances in Information Technology*,
Advances in Intelligent Systems and Computing 266, DOI: 10.1007/978-81-322-1856-2_3,
© Springer India 2014

would be quite impractical in situations where the users are very close to the sensors (e.g. smart house, health care application). Using a Bluetooth or USB, the Wireless Sensor Network can be accessed by a smartphone which is used in everyday life. Thus, use of expensive computers as monitoring unit may be avoided. Not only there are many applications where user mobility is important viz. situations where the sensors and sink nodes are mobile, but also the data collected by various sensor nodes in the wireless sensor network must be made accessible to users at a distant location via Internet. Smartphones may be advantageous here because they have efficient long range communication capabilities, moreover these devices are commonly used by most people.

Standard sensor equipments have limitations in communication, both range wise and cost wise. They are provided with low powered, low range radio transceivers, which uses 802.15.4 standards to conserve the available energy in a Wireless Sensor Node. Also, these radio transceivers are not capable of communicating with conventional communication devices. Conventional communication devices like PDAs, Mobile Phones are provided with WLAN and Bluetooth radio chips for wireless communication, and USB, for serial communication. WLAN is designed to work with conventional Internet system and are more focused on speed and efficiency of communication, not power consumption. Bluetooth consumes less power than WLANs but are designed for Personal Area Network and has only three bit address space hence at most seven devices can connect at a time. So these cannot be implemented directly in Wireless Sensor Networks, hence we propose to use Bluetooth or USB Serial Communication for the communication between Wireless Sensor Networks sink nodes and the user's handheld device.

In [3], a Sensor-Grid infrastructure is presented where sensors along with mobile devices will gather uninterrupted data from monitored objects and users will be facilitated with real-time sensor-data on mobile devices. So the design goal here is to allow direct interaction between the users and the sensor network. Also the user may forward the collected data to the central server for storage. Introducing a mobile device, specifically a smartphone, may be helpful to achieve the aforesaid goal.

In this paper, we describe a cloud based framework which is designed to facilitate communication between the android smartphone and sensors attached to an arduino board. Also we discuss few applications which can benefit from this prototype. Our prototype application is designed for bidirectional communication with the sensor sink node, by listening to the receive buffers for new data on both ends.

Remainder of the paper is organized as follows. In Sect. 2, we discuss related work. The Design and Architecture is illustrated in Sect. 3. We describe the applications that suites this prototype in Sect. 4, It is followed by implementation of our application on the Samsung Galaxy Note with Arduino Board in Sect. 5. Finally we conclude with Sect. 6.

2 Related Work

Using Bluetooth as a gateway between the Wireless Sensor Network and the Internet is implemented in various application with some modifications in each of them. The mobile gateway for remote interaction is proposed in [4]. The idea of Personal Sensor Network and a user centric approach is shown in [5]. The integration of Wireless Sensor Nodes and smartphones in logistic domain is shown in [6]. All these work use Bluetooth as a gateway to the Wireless Sensor Network and considers a cost effective design approach. In the following sections complete design architecture of a Mobile Device as gateway between the Wireless Sensor Network and the Internet is shown, both directions of communication are considered. The cloud user and the Wireless Sensor Node can push or pull data from or to the mobile device.

3 System Architecture

The proposed framework is shown in Fig. 1. Here the smartphone can push and pull data to and from the arduino device. These operation will enable the smartphone user to access the sensor data and to send control information to the sensor network also. The smartphone may use its radio communication services to publish the collected data to the Internet.

The mobile application is set to listen for received data in its serial ports. The arduino firmware is also programmed to listen to its receiver ports for any new data. Thus both devices are ready to receive and send data simultaneously providing a two way communication. In the above said framework, three different kinds of communication technologies are involved.

- 802.15.4 within the WSN
- Bluetooth for handheld devices or OTG cable for USB serial communication
- Cellular network for publishing to the Internet.

So, two points of interconnection among these networks are required. One between WSN and mobile device and another between mobile device and Internet server. Mobile devices have both Bluetooth and cellular radios so it can easily act as a gateway. But sensor motes [7] or nodes are only provided with 802.15.4 radios. All sensor nodes forward data to the sink node using 802.15.4 radios. The sink node is equipped with Bluetooth and USB. The smartphone communicates with the sink using Bluetooth or USB Serial communication.

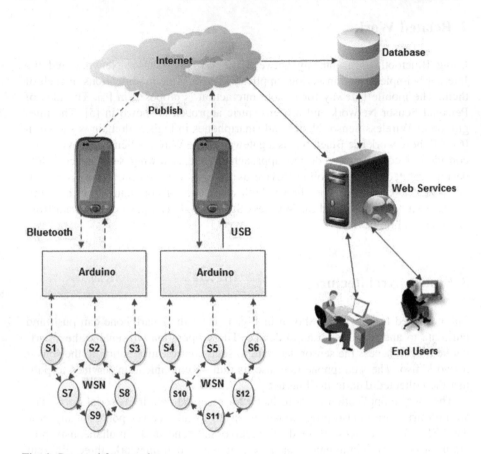

Fig. 1 Proposed framework

3.1 The Arduino Microcontroller

Arduino is an open source hardware board and can be useful in various custom electronic applications. Arduino UNO is used here. It is provided with a AT-Mega328 CPU clocked at 16 MHz and an on-chip flash memory of 32 KB of which 0.5 KB is used by the bootloader. Arduino UNO board [8] provides 14 digital I/O pins and 6 analog pins where sensors can be connected. This makes the arudino board suitable for the design of our application.

4 Applications

Mobile Device as a gateway, may be explored where fixed gateway is not feasible or not accessible physically. Mobile devices may add metadata, to the data collected from sensors before sending to the remote server, making the system cost effective. Few applications are discussed below.

4.1 Example 1: The Agricultural Application

In agricultural system (Fig. 2), this design could be implemented where the soil condition is measured through sensors. Mobile devices which are carried by farmers, can communicate with these sensors to collect data. Thereafter the collected data is sent to the server adding location through its GPS device. Then any person having access to that server can view the soil condition in a detailed map.

4.2 Example 2: Health Care Application

In health care application the sensors are connected to the patient and senses physical parameters of the patient. These collected information is mostly useful to the doctor, hospital and the homecare services. So it is useful if the data can be directly sent to server from which authorized persons can access the data. In Fig. 3, smartphone of the patient is used to collect the sensor data from the sensor devices which are attached on the concerned patient. Also the phone helps in sending the data to the central record management server. The data can also flow in the reverse direction i.e. from the doctors to the sensor or the patients for requesting periodic checkups.

4.3 Example 3: Application in Logistic Domain

Wireless Sensor Nodes sense diverse parameters like shock, tilt, temperature, humidity etc. related to the condition of the transported goods as in Fig. 4. Thus, determine if critical threshold has been reached during the transport. Alert messages can be wirelessly propagated through the Wireless Sensor Network and forwarded to a backend system of logistic provider.

Fig. 2 Agricultural
application

Fig. 3 Health care
application

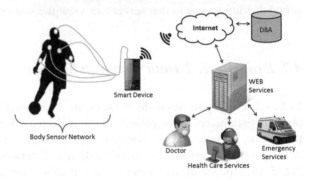

Fig. 4 Logistics application
[6]

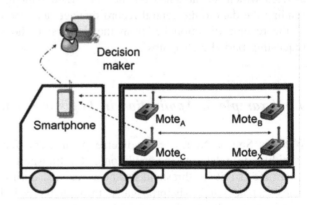

Fig. 5 Implementation set-up

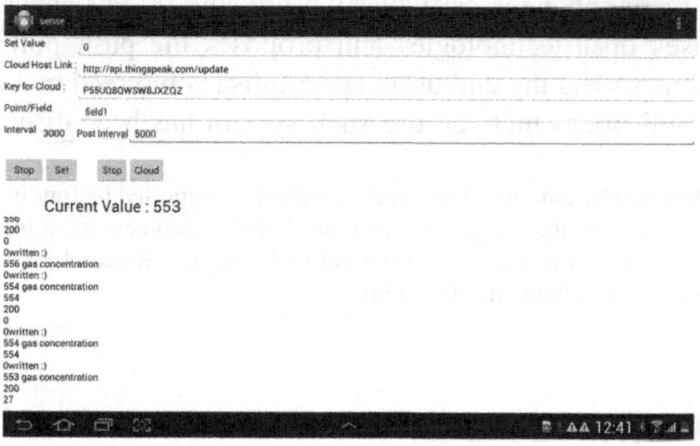

Fig. 6 Android appllication

5 Implementation

In our prototype, we have connected an Arduino UNO board attached with a carbon monoxide sensor, to a Samsung Galaxy Tab via USB OTG cable. The implementation set-up is shown in Fig. 5. The arduino reads the value from the sensor and stores it in its memory. When the arduino receives request from android application for CO concentration it sends the latest CO concentration value from its memory to the android device. After receiving the data, the android device displays the value on its screen and post its value to the registered cloud datastore service, ThingsSpeak [9]. The prototype may facilitate the communication of a sink node of a Wireless Sensor Network with many sensors and a smartphone. For USB Serial communication, Android OS provides Android USB host APIs. In our application, an additional library from google-code is used for detecting devices, requesting permissions and creating end-points for serial transmission.

The android application sends request for data to the arduino board either on the click of a button or based on a timer event. The Arduino board after receiving the request, blinks LED to acknowledge and sends collected sensor-data to the android device. Thus a two way communication between the two devices is accomplished. The smartphone can request arduino to perform a particular function like requesting a particular sensor to send data. After receiving the data the smartphone can take decisive actions according to the result. A screenshot of our application is displayed in Fig. 6.

6 Conclusion

Wireless Sensor Network provide a promising way to enable real-time monitoring of various processes. The monitored data has to be transmitted to the responsible decision makers quickly. For this a long-range connection between the Wireless Sensor Network and the end users are needed. To provide such a connection, use of smartphone is particularly promising. In this paper, the cloud based framework uses open technologies and proposes the push–pull mechanism where both the sensors and the end users accomplishes bidirectional communication. The various applications that can use such system has been discussed here.

Acknowledgments This work is partially supported by funding received from DST-NRDMS for carrying out the research project entitled Development of an Integrated Web portal for Healthcare management based on Sensor-Grid technologies. Research of second author is supported by TCS Research Scholarship Program.

References

1. ITU Report on Internet of Things Executive Summary. www.itu.int/internetofthings/
2. Yick, J., Mukherjee, B., Ghosal, D.: Wireless sensor network survey. Comput. Netw. **52**(12), 2292–2330 (2008)
3. Bhunia, S.S., Roy, S., Mukherjee, N.: On efficient health-care delivery using Sensor-Grid, In: 3rd IEEE International Conference on Emerging Applications of Information Technology (EAIT), pp. 139–142 (2012)
4. Angove, P., O'Grady, M.l., Hayes, J., O'Flynn, B., MP O'Hare, G., Diamond. D.: A mobile gateway for remote interaction with wireless sensor networks. IEEE Sens. J. **11**(12), 3309–3310 (2011)
5. Giorgetti, G., Manes, J., Lewis, H., Mastroianni, S.T., Gupta, SKS.: The personal sensor network: a user-centric monitoring solution. In: Proceedings of the ICST 2nd international conference on body area networks. ICST (Institute for Computer Sciences, Social-Informatics and Telecommunications Engineering), p. 24 (2007)
6. Zller, S., Reinhardt, A., Guckes, H., Schuller, D., Steinmetz, R.: On the Integration of wireless sensor networks and smartphones in the Logistics Domain. In: Proceedings of the 10th GI/ITG KuVS Fachgesprch Drahtlose Sensornetze (FGSN), p. 4952 (2011)
7. TelosB-Wireless measurement system datasheet, Crossbow Technology, Inc., TelosB Mote Platform, Datasheet [Online]
8. Arduino Serial Communication Website. http://arduino.cc/en/reference/serial
9. Thingspeak website. https://www.thingspeak.com/

Design of ECC-Based ElGamal Encryption Scheme Using CL-PKC

Manoj K. Mishra, S. K. Hafizul Islam and G. P. Biswas

Abstract Owing the advantages of Elliptic Curve Cryptography (ECC), we proposed a Certificateless Public Key Cryptosystem (CL-PKC)-based ElGamal Encryption (CL-EE) scheme using ECC in this paper. In literature, ECC-based ElGamal Encryption scheme is devised using Certificate Authority-based Public Key Cryptography (CA-PKC) that needs a global Public Key Infrastructure (PKI) to maintain the public keys and certificates. Moreover, in PKI architecture, the sender must have the additional ability to verify the public key certificate of the receiver. It is known that CL-PKC avoids public key certificate and thus, our CL-EE scheme is more efficient than PKI-based ElGamal Encryption scheme. We simulated our scheme using AVISPA (Automated Validation of Internet Security Protocols and Applications) tool and the results demonstrated that the scheme is secure against both active and passive attacks.

Keywords Elliptic curve cryptography · Certificateless cryptosystem · ElGamal encryption · Public key infrastructure · AVISPA software

M. K. Mishra (✉) · G. P. Biswas
Department of Computer Science and Engineering, Indian School of Mines,
Dhanbad 826004 Jharkhand, India
e-mail: manojmishra.jk@gmail.com

G. P. Biswas
e-mail: gpbiswas@gmail.com

S. K. H. Islam
Department of Computer Science and Information Systems, Birla Institute of Technology
and Science, Pilani 333031 Rajasthan, India
e-mail: hafi786@gmail.com

G. P. Biswas and S. Mukhopadhyay (eds.), *Recent Advances in Information Technology*,
Advances in Intelligent Systems and Computing 266, DOI: 10.1007/978-81-322-1856-2_4,
© Springer India 2014

27

1 Introduction

In 1985, ElGamal [1] proposed ElGamal encryption scheme using CA-PKC that requires PKI to maintain the public keys and the corresponding certificates. The ECC-based [2, 3] ElGamal encryption (PKI-EE) scheme [4] in PKI is shown to be promising over the earlier scheme in terms of security and computation cost. Since ECC-based PKI-EE scheme used elliptic curve point addition and multiplication operations that are much faster than modular exponentiation based on which the earlier scheme was proposed [5]. ECC-based PKI-EE scheme is designed over an elliptic curve cyclic group and its security lies on the difficulties of Decisional Diffie-Hellman (DDH) assumption. However, it suffers from some drawbacks like (1) It needs a PKI to maintain the public keys and certificates, and (2) Users need to verify the corresponding certificate of others for which extra computations are involved. The identity-based cryptosystem (IBC) proposed by Shamir [6] avoids public key certificate since the publicly known identity of a user is used as a public key and the private key is computed and securely handover to the user by a trusted third party, called Private Key Generator (PKG). However IBC-based schemes have one drawback called the private key escrow problem. Since PKG generates the private key of the users, it can easily impersonate any user if PKG is not fully trusted. In 2003, Al-Riyami and Paterson [7] proposed the concept of CL-PKC that removes the certificate management problem of CA-PKC and private key escrow problem of IBC [8–10]. In this paper, we designed the existing PKI-EE scheme in CL-PKC system, called CL-EE. The experimental results of the AVISPA tool [11] shows that our CL-EE scheme is free from active and passive attacks.

The paper is organized as follows. The Sect. 2 discussed about the preliminaries and the Sect. 3 describes the proposed CL-EE scheme. The simulation through the AVISPA tool, security and efficiency discussion of our CL-EE scheme are made in Sect. 4. Finally, Sect. 5 concludes the paper.

2 Preliminaries

2.1 Elliptic Curve Cryptography

Recently, ECC [2, 3] has accepted as an efficient tool in PKC due to the computation, communication and security strengths. For example, it offers same level of security at reduced key sizes than other PKCs. Let E/F_q be a set of elliptic curve points over the prime field F_q, defined by the following equation y^2 mod $q = (x^3 + ax + b)$ mod q, where $x, y, a, b \in F_q$ and $\Delta = (4a^3 + 27b^2)$ mod $q \neq 0$. The additive elliptic curve group defined as $G_q = \{(x, y){:}x, y \in F_q$ and $(x, y) \in E/F_q\} \cup \{O\}$, where the point "$O$" is known as "point at infinity" or "zero point". More about ECC and related descriptions are given in [5].

2.2 Computational Problem

Definition 1 (Computational Diffie-Hellman Problem (CDHP)). Given $\langle P, aP, bP \rangle$ for any $a, b \in Z_q^*$, computation of abP is hard to the group G_q.

Definition 2 (Decisional Diffie-Hellman Problem (DDHP)). Given $\langle P, aP, bP, cP \rangle$ for any $a, b, c \in Z_q^*$, decide whether $cP = abP$ i.e., $c = ab \bmod q$.

2.3 Message-To-Point Conversion

Koblitz [12] introduced a probabilistic technique that converts a message to an elliptic curve point and vice versa. Assume that m is a message and k be a large integer such that $0 \le m < M$ and $q > Mk$. Thus given m and for each $j = 1, 2, ..., k$ we obtain $x \in F_q$, corresponding to $mk + j$. For such an x, we compute $f(x) = x^3 + ax + b$ and try to find a square root of $f(x)$. If a y is found such that $y^2 = f(x)$, we take $P_m = (x, y)$. If not, we incremented j by 1 and try again with the corresponding x until an x is found for which $f(x)$ is a square before j gets bigger than k. The message m can also be recovered from P_m using $m = [(\bar{x} - 1)/k]$, where $\bar{x}$ is the integer corresponding to x.

3 Proposed CL-EE Scheme

- **Setup:** For given a security parameter $k \in Z^+$, PKG executes this algorithm only once to establish the system's parameter Ω in the following way.

  (a) Choose a k-bit prime q and determine the tuple $\langle F_q, E/E_q, G_q, P \rangle$.
  (b) Choose $x \in Z_q^*$ as private key and $P_{pub} = xP$ as public key.
  (c) Choose a one-way secure hash functions $H:\{0, 1\}^* \to \{0, 1\}^k$.
  (d) Publish $\Omega = \langle F_q, E/F_q, G_q, P, P_{pub}, H \rangle$ as system's parameter.

- **Set-Secret-Value:** User $ID_i, i \in \{A, B\}$ selects a $x_i \in_R Z_q^*$ as his secret value and computes the corresponding public key as $P_i = x_iP$.
- **Partial-Private-Key-Extract:** To get the partial private key, user ID_i sends $\langle ID_i, P_i \rangle$ to PKG and then PKG does as follows:

  (a) Select a $r_i \in_R Z_q^*$ and compute $R_i = r_iP$.
  (b) Compute $d_i = r_i + xH(ID_i, R_i, P_i) \bmod q$.

Now PKG sends $\langle d_i, R_i \rangle$ via a secure channel to ID_i. The identity-based public key of ID_i is $Q_i = R_i + H(ID_i, R_i, P_i)P_{pub}$ and the private/public key pair $\langle d_i, Q_i \rangle$ can be verified by checking whether the equation $Q_i = R_i + H(ID_i, R_i, P_i)P_{pub} = d_iP$ holds. Since we have

$$R_i + H(ID_i, R_i, P_i)P_{pub} = r_iP + H(ID_i, R_i, P_i)xP$$
$$= (r_i + H(ID_i, R_i, P_i)x)P$$
$$= (r_i + h_ix)P$$
$$= d_iP$$
$$= Q_i$$

- **Set-Private-Key:** User ID_i takes the pair $sk_i = \langle d_i, x_i \rangle$ as full private key.
- **Set-Public-Key:** User ID_i takes $pk_i = \langle P_i, R_i \rangle$ as full public key.
- **CL-EE-Enc:** To encrypt a message m, Sender ID_A uses the full public key $pk_B = \langle R_B, P_B \rangle$ of the Receiver ID_B and computes the following:

  (a) Convert the message m to an elliptic curve point $P_m \in G_q$.
  (b) Choose a number $t \in {}_R Z_q^*$ and compute $C_1 = tP$.
  (c) Compute $C_2 = P_m + t(P_B + R_B + H(ID_B, R_B, P_B)P_{pub})$.
  (d) Output $\langle C_1, C_2 \rangle$ as the ciphertext of the message m.

- **CL-EE-Dec:** To decrypt $\langle C_1, C_2 \rangle$, Receiver ID_B users his own full private key $sk_B = \langle d_B, x_B \rangle$ and performs the following:

  (a) Compute $P_m = C_2 - (x_B + d_B)C_1$
  (b) Convert the point P_m back to the original message m.

- **Correctness:** Since we have

$$C_2 - (x_B + d_B)C_1 = P_m + t\big(P_B + R_B + H(ID_B, R_B, P_B)P_{pub}\big) - (x_B + d_B)tP$$
$$= P_m + t(x_BP + r_BP + h_BxP) - (x_B + d_B)tP$$
$$= P_m + t(x_B + r_B + h_Bx)P - (x_B + d_B)tP$$
$$= P_m + t(x_B + d_B)P - t(x_B + d_B)P$$
$$= P_m$$

Thus, the proposed CL-EE scheme is correct.

4 Analysis of the Proposed CL-EE Scheme

4.1 Simulation of the Proposed CL-EE Scheme Using AVISPA Tool

The formal verification of the proposed CL-EE scheme using AVISPA software have been done in this section. We implemented our scheme (See Fig. 1a, b) in AVISPA tool [13, 14] using a role based language called HLPSL (High Level Protocol Specification Language).

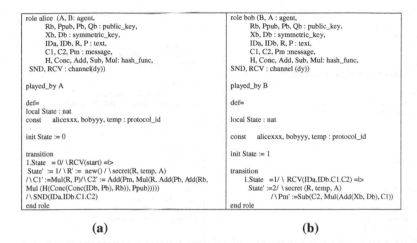

role alice (A, B: agent,	role bob (B, A : agent,
Rb, Ppub, Pb, Qb : public_key,	Rb, Ppub, Pb, Qb : public_key,
Xb, Db : symmetric_key,	Xb, Db : symmetric_key,
IDa, IDb, R, P : text,	IDa, IDb, R, P : text,
C1, C2, Pm : message,	C1, C2, Pm :message,
H, Conc, Add, Sub, Mul: hash_func,	H, Conc, Add, Sub, Mul: hash_func,
SND, RCV : channel(dy))	SND, RCV : channel (dy))
played_by A	played_by B
def=	def=
local State : nat	
const alicexxx, bobyyy, temp : protocol_id	local State : nat
init State := 0	const alicexxx, bobyyy, temp : protocol_id
transition	init State := 1
1.State = 0/ \RCV(start) =|>	
State' := 1/ \R' := new() / \ secret(R, temp, A)	transition
/\ C1' :=Mul(R, P)/\ C2' := Add(Pm, Mul(R, Add(Pb, Add(Rb,	1.State =1/ \ RCV(IDa.IDb.C1.C2) =|>
Mul (H(Conc(Conc(IDb, Pb), Rb))), Ppub)))))	State' :=2/ \ secret (R, temp, A)
/\ SND(IDa.IDb.C1.C2)	/\ Pm' :=Sub(C2, Mul(Add(Xb, Db), C1))
end role	end role

(a)   (b)

Fig. 1 Role specification of the **a** Sender and **b** Receiver in HLPSL

AVISPA supports four model checkers (1) On-the-fly Model-Checker (OFMC) that perform several symbolic techniques to explore the state space in a demand-driven way, (2) CL-AtSe (Constraint-Logic-based Attack Searcher) which translate a security protocol specification into a set of constraints, (3) SATMC (SAT-based Model-Checker) that explores the state space through several symbolic techniques and (4) TA4SP (Tree Automata based on Automatic Approximations for the Analysis of Security Protocols) approximate the intruder knowledge by using the propositional formula and regular tree languages. Our CL-EE scheme is simulated using the SPAN (Security Protocol ANimator) for AVISPA. The simulation results of OFMC, CL-AtSe, SATMC and TA4SP (see Figs. 2a, b, 3a, b) model checker informs that our scheme is secure from passive and active attacks.

4.2 Security and Efficiency Analysis of the Proposed CL-EE Scheme

According to [7], two types of adversaries $\mathcal{A}_I$ and $\mathcal{A}_{II}$ are present in any CL-PKC with different capabilities. The Type I adversary $\mathcal{A}_I$ (dishonest user) cannot access PKG's secret key but, he can replace the public key of any user with a value chosen by him. The Type II adversary $\mathcal{A}_{II}$ (malicious PKG) can access PKG's master secret key but, he cannot replace the public key of any user.

(1) The adversary $\mathcal{A}_I$ have knowledge about the secret value x'_B corresponding to the replaced public key $P'_B = x'_B P$ but, he has no access to identity-based private key d_B of ID_B. To forge our CL-EE scheme, $\mathcal{A}_I$ has to solve the DDH problem from $(P, C_1 = tP, Q_B = d_B P)$. Thus, the security of our CL-EE scheme is based on the DDH assumption against $\mathcal{A}_I$.

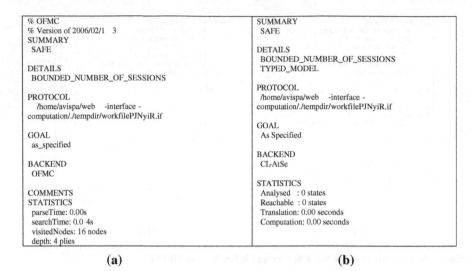

```
% OFMC                                    SUMMARY
% Version of 2006/02/1  3                   SAFE
SUMMARY
  SAFE                                    DETAILS
                                            BOUNDED_NUMBER_OF_SESSIONS
DETAILS                                     TYPED_MODEL
  BOUNDED_NUMBER_OF_SESSIONS
                                          PROTOCOL
PROTOCOL                                    /home/avispa/web   -interface -
  /home/avispa/web   -interface -         computation/./tempdir/workfilePJNyiR.if
computation/./tempdir/workfilePJNyiR.if
                                          GOAL
GOAL                                        As Specified
  as_specified
                                          BACKEND
BACKEND                                     CL-AtSe
  OFMC
                                          STATISTICS
COMMENTS                                    Analysed  : 0 states
STATISTICS                                  Reachable : 0 states
  parseTime: 0.00s                          Translation: 0.00 seconds
  searchTime: 0.0 4s                        Computation: 0.00 seconds
  visitedNodes: 16 nodes
  depth: 4 plies
```

(a)   (b)

Fig. 2 Simulation results in **a** OFMC and **b** CL-AtSe model checkers

```
SUMMARY                                   SUMMARY
  SAFE                                      SAFE

DETAILS                                   DETAILS
  STRONGLY_TYPED_MODEL                      TYPED_MODEL
  BOUNDED_NUMBER_OF_SESSIONS                OVER_APPROXIMATION
  BOUNDED_SEARCH_DEPTH                      UNBOUNDED_NUMBER_OF_SESSIONS
  BOUNDED_MESSAGE_DEPTH
                                          PROTOCOL
PROTOCOL               \                    /home/avispa/web  -interface -computation
  workfilePJNyiR.if                       /./tempdir/workfile7hHHIS.if.ta4sp

GOAL                                      GOAL
  See the HLPSL specification..             SECRECY As specified in HLPSL/IF

BACKEND                                   BACKEND
  SATMC                                     TA4SP

COMMENTS                                  COMMENTS
                                            TA4SP uses abstractions '2AgentsOnly'
STATISTICS                                  For the given i nitial state, an over -approximation is used
  attackFound        false    boolean      with an unbounded number of sessions.
  upperBoundReached  true     boolean      Terms supposed not to be known by the intruder are still secret.
  graphLeveledOff    2        steps
  satSolver          zchaff   solver      STATISTICS
  maxStepsNumber     11       steps         Translation: 0.01 seconds
  stepsNumber        2        steps         Computation 0.36 seconds
  atomsNumber        0        atoms
  clausesNumber      0        clauses
  encodingTime       0.0      seconds     ATTACK TRACE
  solvingTime        0        seconds       No attack found
  if2sateCompilationTime 0.07 seconds

ATTACK TRACE
  No attacks have been found..
```

(a)   (b)

Fig. 3 Simulation results in **a** SATMC and **b** TA4SP model checkers

(2) The adversary $\mathcal{A}_{II}$ knows the identity-based private key d_B of ID_B but, the public key $P_B = x_B P$ of ID_B cannot be replaced by a value chosen by him. To forge our CL-EE scheme, $\mathcal{A}_{II}$ tries to solve the DDH problem from $\langle P, C_1 = tP, P_B = x_B P \rangle$ that is assumed to be hard. Thus, our CL-EE scheme is secure based on DDH assumption against Type II adversary $\mathcal{A}_{II}$.

The PKI-EE scheme [4] executes $\langle C_1, C_2 \rangle = \langle tP, P_m + tP_B \rangle$ for encryption and $P_m = C_2 - x_B C_1$ for decryption, where $\langle x_B, Y_B = x_B P \rangle$ is the private/public key pair of ID_B. Thus, the computation cost of PKI-EE scheme is $3T_{EM} + 2T_{EA}$, where T_{EM} and T_{EA} are time complexities of an elliptic curve point multiplication and addition. In our scheme, the computation $P_B + R_B + H(ID_B, R_B, P_B)P_{pub}$ can be done in off-line mode and thus the computation cost is $3T_{EM} + 2T_{EA}$.

5 Conclusion

The authors have proposed an ECC-based ElGamal algorithm based on CL-PKC, called CL-EE and simulated it in the AVISPA tool for formal security validation. Similar to the PKI-EE scheme, the security of the proposed scheme is also based on the DDH assumption against the adversaries $\mathcal{A}_I$ and $\mathcal{A}_{II}$. Although, our scheme and PKI-EE scheme have same computation overhead however, public key certificate is removed in our CL-EE scheme.

References

1. ElGamal, T.: A public key cryptosystem and a signature protocol based on discrete logarithms. IEEE Trans. Info. Theor. **31**, 469–472 (1985)
2. Miller, V.S.: Use of elliptic curves in cryptography. In: Proceeding of the Advances in Cryptology (Crypto'85), pp. 417–426. Springer, Berlin (1985)
3. Koblitz, N.: Elliptic curve cryptosystem. J. Math. Comp. **48**(177), 203–209 (1987)
4. Stallings, W.: Cryptography and Network Security Principles and Practice, pp. 325–326. Prentice Hall (Pearson), Upper Saddle River (2006)
5. Hankerson, D., Menezes, A., Vanstone, S.: Guide to elliptic curve cryptography. Springer, New York (2004)
6. Shamir, A.: Identity based cryptosystems and signature schemes. In: Proceedings of the Crypto'84, LNCS, vol. 196, pp. 47–53. Springer, Berlin (1984)
7. Al-Riyami, S., Paterson, K.: Certificateless public key cryptography. In: Proceedings of the Asiacrypt'03, LNCS, vol. 2894, pp. 452–473. Springer, Berlin (2003)
8. Islam, S.H., Biswas, G.P.: Certificateless short sequential and broadcast multisignature schemes using elliptic curve bilinear pairings. J. King Saud Univ.-Comput. Inf. Sci. **26**(1), 89–97 (2014)
9. Islam S.H., Biswas G.P.: Certificateless Strong Designated Verifier Multisignature Scheme Using Bilinear Pairings. In: Proceedings of the International Conference on Advances in Computing, Communications and Informatics (ICACCI'12), pp. 540–546. (2013)

10. Islam, S.H., Biswas, G.P.: Provably secure certificateless strong designated verifier signature scheme based on elliptic curve bilinear pairings. J. King Saud Univ.-Comput. Inf. Sci. **25**, 51–61 (2013)
11. AVISPA Web tool: Automated Validation of Internet Security Protocols and Applications. www.avispa-project.org/web-interface/. Accessed on July 2013
12. Koblitz, N.: A Course in Number Theory and Cryptography. Springer, Berlin (1991)
13. Islam, S.H., Biswas, G.P.: An efficient and secure strong designated verifier signature scheme without bilinear pairings. J. Appl. Math. Info. **31**(3–4), 425–441 (2013)
14. Islam, S.H., Biswas, G.P.: A provably secure identity-based strong designated verifier proxy signature scheme from bilinear pairings. J. King Saud Univ.-Comput. Inf. Sci. **26**(1), 55–67 (2014)

Part II
Technical Session-II: Image Processing Applications

A Fast Digital-Geometric Approach for Granulometric Image Analysis

Sahadev Bera, Arindam Biswas and Bhargab B. Bhattacharya

Abstract A simple algorithm for automated analysis of granulometric images consisting of touching or overlapping convex objects such as coffee bean, food grain, is presented. The algorithm is based on certain underlying digital-geometric features embedded in their snapshots. Using the concept of an outer isothetic cover and geometric convexity, the separator of two overlapping objects is identified. The objects can then be isolated by removing the isothetic covers and the separator. The technique needs only integer computation and its termination time can be controlled by choosing a resolution parameter. Experimental results on coffee beans and other images demonstrate the efficiency and robustness of the proposed method compared to earlier watershed-based algorithms.

Keywords Granulometric image analysis · Outer isothetic cover · Digital geometry · Coffee bean segmentation

1 Introduction

Fast and automated analysis of a granulometric image consisting of convex objects such as agricultural products, coffee beans, nuts, cookies, and chocolates, has many practical applications in the field of digital image segmentation. Several

S. Bera (✉) · B. B. Bhattacharya
Advanced Computing and Microelectronics Unit, Indian Statistical Institute, Kolkata, India
e-mail: sahadevbera@gmail.com

B. B. Bhattacharya
e-mail: bhargab@isical.ac.in

A. Biswas
Department of Information Technology, Bengal Engineering and Science University, Shibpur, Howrah, India
e-mail: barindam@gmail.com

G. P. Biswas and S. Mukhopadhyay (eds.), *Recent Advances in Information Technology*,   37
Advances in Intelligent Systems and Computing 266, DOI: 10.1007/978-81-322-1856-2_5,
© Springer India 2014

techniques have been proposed for agricultural product inspection [10] using X-ray images. X-ray images provide the internal product details, which allow the analysts to detect the presence of damage due to worms, or other defects by non-destructive (non-invasive) methods; worms contribute to conditions favoring mold growth and toxin production. There exist various works on segmentation of agricultural product [3, 18]. Most of them are based on watershed transform [5, 18, 19], mathematical morphology [7], granulometric methods [20], or variants.

Among the various image segmentation techniques, the watershed algorithm is a popular segmentation method, which originates from the concept of mathematical morphology [19]. This technique has been successfully applied for gray-tone image segmentation in various fields including medicine [2], computer vision [15], biomedicine [4], signal processing [12], industry [13], remote sensing [9], computer-aided design [16], and video coding [21]. The watershed algorithm has also been applied for colored image segmentation [8].

When two or more objects in a binary image overlap or touch each other, a single connected object is formed. In order to analyze different characteristics of the objects, it is necessary to segment them into individual components. In this regard, a well known technique is the morphological watershed algorithm, which uses distance transforms [6, 14]. The watershed algorithm segments an image into different regions by treating its inverse distance map as a landscape and the local minima as markers. Each of the segmented regions is labeled with a unique index. Different objects can be separated and identified using the indices of the segmented regions. The effective performance of watershed segmentation depends on the selection of local minima or markers. The spurious markers lead to over-segmentation, which is a major drawback of the watershed algorithm. The performance becomes worse when the objects are irregular-shaped, overlapped or connected, as more spurious local minima tend to occur in the distance transform. Thus, a preprocessing of the markers is needed to improve the performance.

This paper presents a new algorithm for the segmentation of touching or overlapping convex [11] objects based on digital geometry. We use the concept of outer isothetic cover (OIC) [1] to determine the joining points of the edges of two objects. Next, the matching pair of two joining points is formed using the convexity property. The straight line segment joining a pair indicates the separator of two overlapping objects. The isolated objects are obtained by removing the isothetic covers and the straight line segments. The advantage of the proposed method lies in the fact that the over-segmentation error is significantly reduced and the required computation is limited to the integer domain only. The rest of the paper is organized as follows. The preliminaries are given in Sect. 2. The proposed method is described in Sect. 3. A demonstration is shown in Sect. 4. Comparative results with the watershed segmentation technique are presented in Sect. 5. Finally, conclusions are drawn in Sect. 6.

2 Definitions

The definition of digital convexity [11] is given as follows.

Definition 1. *A finite set $M \in \mathbb{Z}^2$ is digitally convex if and only if either of the following is true:*

1. *For all $< p, q > \in M$, at least one digital straight segment (DSS) that has p and q as end pixels is contained in M.*
2. *For all $< p, q, r > \in M$, all of the grid points in the Δpqr are in M.*

3 Granulometric Segmentation Based on Digital Geometry

The proposed segmentation method based on digital geometry consists of the following steps.

3.1 Deriving the Outer Isothetic Cover

Let an input binary image $\mathcal{I}$ be given, which consists of several overlapped convex objects. The outer isothetic cover (OIC) [1], which is the minimum-area isothetic polygon perceived on a background grid enclosing $\mathcal{I}$, is first determined. It may be noted that the OIC can be computed for different grid sizes, $g = 1, 2, 3, \ldots$, where g (resolution parameter) indicates the horizontal and vertical spacings of the background grid. When $g = 1$, the most accurate approximation of the geometrical shape of the object is captured (highest resolution). Hence, in the proposed algorithm the OIC for $g = 1$ is used for obtaining the best segmentation results. Note that the OIC provides the tightest outer approximation of the object that does not introduce any concavity; rather it removes the smaller concavities from the boundary of the object. The convexity of an object is a geometrical shape information, which is inherited in its OIC, and hence, we have the following lemma.

Lemma 1. *The OIC of a digitally convex object is also digitally convex.*

Proof. Assume that the OIC, S, of a convex digital object, C, is not convex. Then S must contain at least one instance of two consecutive 270° vertices. Thus the four object occupying cells should be arranged in the form of a 'U'-turn. Therefore, the object is no longer digitally convex. Hence, the proof follows by contradiction.

The OIC of a concave object may not be necessarily concave; it depends on the size and the location of the concavity and the grid size g of the OIC. The small concavities of an object may not be reflected in its OIC at higher grid sizes. However, at lower grid size, the concavities of the object may also be present in the OIC. In Fig. 1a, the OIC of a concave object appears as concave, whereas in

Fig. 1 OIC of concave
objects

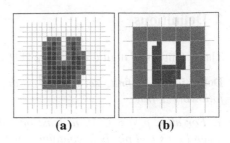

(a)   (b)

Fig. 1b, the OIC of the same object becomes convex at larger grid size. Thus, we
have the following lemmas.

Lemma 2. *The OIC of a concave object may or may not be concave.*

Lemma 3. *If the OIC is concave then the object is concave. However, if the OIC
is convex then the object can either be convex or concave.*

Note that if two convex objects overlap, then the resultant object may be either
convex or concave. But when two or more circular/elliptic-shaped objects overlap
then the resultant object is concave.

3.2 Identification and Management of Joining Points

A granulometric image $\mathcal{I}$ often contains circular/elliptic-shaped (convex) overlapping objects. When two such convex objects overlap partially, they give rise to
two corresponding concavities at the resultant boundary. Since the OIC of a
convex object is also digitally convex, it will not contain two consecutive 270°
vertices. Let the points, where one of them occludes the other be termed as *joining
points*. Then, in the resultant OIC of the two overlapping convex objects, the
concavities, i.e., the *joining points* will be indicated by two consecutive 270°
vertices (shown in Fig. 2a). The midpoint of two consecutive 270° vertices is
considered as the *joining point*. Also, a joining point may occur when the two sides
defining a 270° vertex are long (as shown in Fig. 2b). In this case, the 270° vertex
is considered as the joining point. When two or more circular/elliptic objects
overlap with each other, the joining points are detected in the above-mentioned
manner.

The joining points, thus obtained, are stored in a linked list $\mathcal{L}$ as the OIC is
traversed along its boundary. Each node contains the coordinates of the joining
point, (x, y), pointers to previous and next joining points, and a flag d. The value of
the flag, $d = 1$, indicates that the joining point occurs on the OIC, that describes
the outer border (called *border OIC*), and $d = 0$ indicates that the joining point
belongs to the OIC forming a hole (called *hole OIC*) inside the overlapping
objects. For each separate OIC, such a linked-list is formed, and finally, the roots
of these lists are stored in another list.

Fig. 2 Outer isothetic cover of two convex objects **a** having consecutive 270° vertices **b** having one 270° vertex with long edge

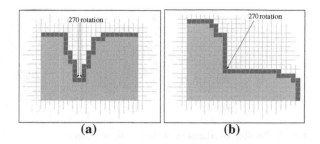

(a)   (b)

3.3 Determining the Matching Pairs and the Separators

In order to identify the lines of separation (called *separators*) for two overlapping objects, the joining points are to be matched to form suitable matching pairs. If a connected component of the image consists of only two overlapping objects, there will be two joining points, and the separator is obtained by joining these two points as shown in Fig. 3a.

For a component with three overlapping objects, several combinations are possible:

(a) Objects (a_1, a_2, a_3) overlap linearly, i.e., a_1 overlaps with a_2 and a_2 with a_3; here, four joining points are detected. In this case, two adjacent joining points, where the concavities of the OIC are in opposite direction, form the matching pair. The direction of a concavity at a joining point of the OIC may be determined by the orientation of its two consecutive 270° vertices.

Figure 3b shows such a scenario, whether the joining points are shown in blue and the separators are shown with green lines.

(b) When three objects overlap on each other such that there is no hole in their interior, only three joining points (shown in blue dots) will appear. A dummy joining point (red dot) is introduced at the centroid of these joining points (Fig. 3c). The dummy joining point is paired with each of the three joining points.

(c) The three objects overlap with each other, but they create a hole at the interior (Fig. 3d). There are two OICs, one outer border OIC ($d = 1$) and the other OIC is a hole ($d = 0$); both of these generate three joining points each. Note that, in the presence of a hole OIC, no two joining points on the outer border cover will form a matching pair. Each joining point of the hole OIC is matched with the nearest joining point of the border OIC.

Figure 3d shows the hole joining points (yellow dots), the border joining points (blue dots), and the three separators.

Lemma 4. *No two joining points on the hole OIC will form a matching pair.*

Proof. Let p and q be the two joining points lying on the same hole OIC, which form a matching pair. Then the line segment joining them, which divides the inner

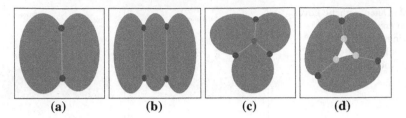

Fig. 3 Possible overlapping of two or three objects

hole, is of no significance, as a separator is intended to distinguish two overlapping objects. Hence the proof.

Similarly, when multiple objects overlap, each of the joining points on the hole OIC is matched with its nearest joining point on the border OIC. All the pairs of joining points are finally connected by straight line segments to determine the separators.

3.4 Segmenting the Objects

The collection of the separators partitions the input image into several regions, where each region corresponds to a convex object. The total count of pixels in each object gives the approximate area of the object. Also, the shape of these regions may be used to analyze the granulometric attributes of the objects.

4 Demonstration of the Proposed Method

A step-wise demonstration of the proposed method on a coffee bean image is shown in Fig. 4. The input coffee bean image shown in Fig. 4a contains a numbers of beans colored as black. The outer isothetic covers (both hole OIC and border OIC indicated in red) are shown in Fig. 4b. The joining points are marked (shown with blue dots) when two consecutive 270° vertices, or a 270° vertex with long arms, are observed while deriving the OICs (Fig. 4c). Each of the joining points of the hole OIC is paired with its nearest joining point on the border OIC; a separator is drawn for each matching pair (Fig. 4d). The remaining joining points are paired with the nearest unpaired joining point and the separators are drawn as shown in Fig. 4e. Figure 4f shows the original image segmented into its components obtained after removing the isothetic cover and the separating line segments.

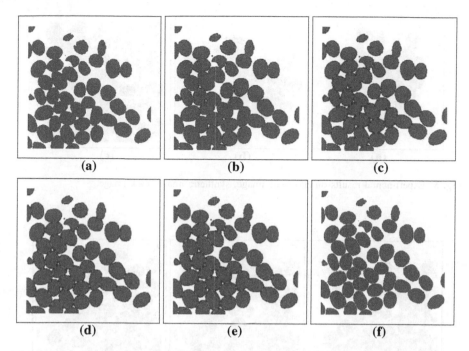

Fig. 4 Step-wise snapshots of the experiment on a coffee bean image: **a** input image **b** after drawing outer isothetic cover **c** after finding the joining points **d** after joining the matching pair for points on OIC with $d = 0$ **e** drawing the other separating lines **f** final result of segmentation

5 Experimental Results

The proposed algorithm has been implemented in Matlab version R2011b on the openSUSE^{TM} OS Release 12.3 HP Compaq with Intel$^{\circledR}$ Pentium Dual-Core, 1.7 GHz processor. The algorithm has been tested on several images. The input image [5] shown in Fig. 4a consists of 47 coffee beans, in which some of them are broken. The segmented image obtained by the proposed algorithm is shown in Fig 4f with 47 separated coffee beans. We also include the results on several other images [5, 17] as shown in Fig. 5a, c, e.

We have also run a watershed-based algorithm on these examples. Figure 6 shows the segmentation results on these images with different distance transforms (DT) used in the watershed segmentation method. We have implemented three distance transformations, namely Euclidean, city block, and chessboard [5] for the watershed algorithm (see Table 1), two such sample input images are shown in

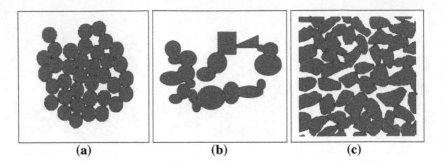

Fig. 5 Experimental results on chocolate image, synthetic image, rock image

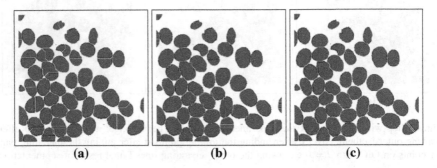

Fig. 6 Results of watershed segmentation using three distance transformations (DT) on coffee beans image **a** Euclidean **b** City block **c** Chessboard

Fig. 7. In most of the cases, we observe that they produce over-segmentation. The proposed algorithm is run with the two values of the resolution parameter ($g = 1$, 2). For $g = 1$, our algorithm produces an exact segmentation. For $g = 2$, the execution time is reduced significantly compared to that of the watershed-based algorithms. However, this may produce a slight under-segmentation in some cases.

Table 1 Comparative results for images

Image	Size	NOP	NO	Watershed algorithm using						Proposed algorithm			
				Euclidean DT		City block DT		Chessboard DT		OIC (g = 1)		OIC (g = 2)	
				NDO	T	NDO	T	NDO	T	NDO	T	NDO	T
Figure 4a	278 × 276	29573	47	65	0.0895	48	0.1041	48	0.1129	47	0.1482	46	0.0437
Figure 5a	423 × 411	70397	36	41	0.2054	40	0.2423	41	0.2689	36	0.3329	35	0.1015
Figure 5c	390 × 570	87375	21	45	0.2612	23	0.3029	18	0.3302	21	0.4692	20	0.1443
Figure 7a	1352 × 1960	630764	83	171	3.0463	85	3.5599	86	3.7713	83	5.6303	83	1.9316
Figure 7b	1352 × 1960	667830	42	123	3.0978	42	3.5666	42	3.8108	42	5.5587	42	1.9133

NOP Number of object pixels; *NO* Number of objects; *NDO* Number of detected objects; *T* CPU time in seconds

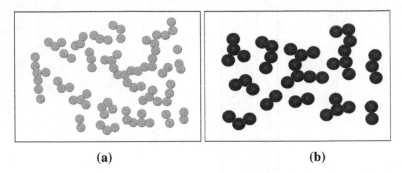

(a)   (b)

Fig. 7 Experiment on different images **a** biscuits **b** sweets

6 Conclusion

In this paper, we have described an automated technique for granulometric object segmentation based on digital geometric features of the underlying binary image. It needs only integer domain computation and outperforms the classical watershed-based algorithms in performance.

References

1. Biswas, A., Bhowmick, P., Bhattacharya, B.B.: Construction of isothetic covers of a digital object: a combinatorial approach. J. Vis. Commun. Image Represent. **21**(4), 295–310 (2010)
2. Cates, J.E., Whitaker, R.T., Jones, G.M.: Case study: an evaluation of user-assisted hierarchical watershed segmentation. Med. Image Anal. **9**, 566–578 (2005)
3. Casasent, D., Talukdar, A., Cox, W., Chang, H., Weber, D.: Detection segmentation and pose estimation of multiple touching product inspection items. In Meye, G., DeShazer, J. (eds.) Optics in Agriculture, Forestry, and Biological Processing II, vol. 2907, pp. 205–216 (1996)
4. Charles, J.J., Kuncheva, L.I., Wells, B., Lim, I.S.: Object segmentation within microscope images of palynofacies. Comput. Geosci. **34**, 688–698 (2008)
5. Chen, Q., Yang, X., Petriuchen, E.M.: Watershed segmentation for binary images with different distance transforms. In proceedings HAVE, pp. 111–116 (2004)
6. Dougherty, E.R.: An Introduction to Morphological Image Processing. SPIE Optical Engineering Press, Washington (1992)
7. Iwanowski, M.: Morphological boundary pixel classification. In proceedings EUROCON, pp. 146–150 (2007)
8. Jung, C.R.: Unsupervised multiscale segmentation of color images. Pattern Recognit. Lett. **28**, 523–533 (2007)
9. Karantzalos, K., Argialas, D.: Improving edge detection and watershed segmentation with anisotropic diffusion and morphological levellings. Int. J. Remote Sens. **27**(24), 5427–5434 (2006)
10. Keagy, P.M., Parvin, B., Schatzki, T.F.: Machine recognition of navel worm damage in X-ray images of pistachio nuts. Lebensm-Wiss U Technol **29**, 140–145 (1996)

11. Klette, R., Rosenfeld, A.: Digital geometry: geometric methods for digital picture analysis. Morgan Kaufmann series in computer graphics and geometric modeling. Morgan Kaufmann, San Francisco (2004)
12. Leprettre, B., Martin, N.: Extraction of pertinent subsets from time–frequency representations for detection and recognition purposes. Signal Process. **82**, 229–238 (2002)
13. Malcolm, A.A., Leong, H.Y., Spowage, A.C., Shacklock, A.P.: Image segmentation and analysis for porosity measurement. J. Mater. Process. Tech. **192–193**, 391–396 (2007)
14. Orbert, C.L., Bengtsson, E.W., Nordin, B.G.: Watershed segmentation of binary images using distance transformations. In: Proceedings of SPIE, vol. 1902, pp. 159–170 (1993)
15. Park, S.C., Lim, S.H., Sin, B.K., Lee, S.W.: Tracking non-rigid objects using probabilistic Hausdorff distance matching. Pattern Recognit. **38**, 2373–2384 (2005)
16. Razdan, A., Bae, M.S.: A hybrid approach to feature segmentation of triangle meshes. Comput. Aided Des. **35**, 783–789 (2003)
17. Sun, H.Q., Luo, Y.J.: Adaptive watershed segmentation of binary particle image. J. Microsc. **233**(2), 326–330 (2009)
18. Talukder, A., Casasent, D., Lee, H., Keagy, P.M., Schatzki, T.F.: Modified binary watershed algorithm for segmentation of X-ray agricultural products. In Proceedings of SPIE, vol. 3543 (1998)
19. Vincent, L., Soille, P.: Watersheds in digital spaces: an efficient algorithm based on immersion simulations. IEEE Trans. Pattern Anal. Mach. Intell. **13**(6), 583–598 (1991)
20. Vincent, L.: Fast granulometric methods for the extraction of global image information. In proceedings of PRASA, pp. 119–140. Broederstroom, South Africa (2000)
21. Wang, D.: Unsupervised video segmentation based on watersheds and temporal tracking. IEEE Trans. Circ. Syst. Video Technol. **8**(5), 539–546 (1998)

Word-Level Handwritten Script Identification from Multi-script Documents

Mallikarjun Hangarge and K. C. Santosh

Abstract In this paper, we present the directional discrete cosine transform (DCT) based rotation invariant features for word-level handwritten script identification. Our aim in this paper is two folds: one is to validate the effectiveness of the directional DCT (D-DCT) in extracting edge information of the studied word image and another is to provide rotation invariant property since conventional DCT (C-DCT) does not offer both issues. For each extracted word image, we compute DCT, its coefficient matrix and decompose into different directions such as horizontal, vertical, left and right diagonals plus mean and standard deviations of the decomposed components. These statistical features are then evaluated with hundreds of word images from six different scripts by using linear discriminant analysis (LDA) and achieved an accuracy of 97.35 % in average.

Keywords Discrete cosine transform · Rotation invariant features · Script identification · Multi-script document

1 Introduction

In document image processing, multi-script handwritten character recognition has been received an important attention since a few decades. Hitherto, no multi-script OCR exists to handle real-world documents with several different scripts, more specifically Indic handwritten documents where various scripts can be found in

M. Hangarge (✉)
Karnatak Arts, Science and Commerce College, Bidar, Karnataka, India
e-mail: mhangarge@yahoo.co.in

K. C. Santosh (✉)
Communications Engineering Branch, US National Library of Medicine National Institutes of Health, Bethesda, MD, USA
e-mail: santosh.kc@nih.gov

G. P. Biswas and S. Mukhopadhyay (eds.), *Recent Advances in Information Technology*, 49
Advances in Intelligent Systems and Computing 266, DOI: 10.1007/978-81-322-1856-2_6,
© Springer India 2014

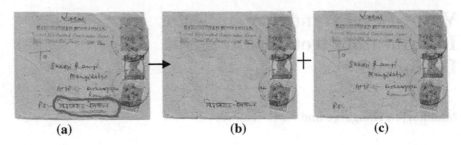

Fig. 1 An example showing the script identification, where **a** a bi-script sample is expected to separate **b** Devanagari and **c** English script

word, line and sometimes paragraph levels. In this context, an automatic identification or separation of different script zones from a multi-script document basically enhances the OCR performance. And of course, an automatic script separation facilitates multi-lingual document indexing, sorting, retrieval and further knowledge discovery. Considering the difficulties and challenging issues associated with Indic handwritten multi-script document processing, in this paper, we attempt a generic method for script identification and thus we aim to apply it as a precursor to OCR. Figure 1 shows the expected outcomes of the input bi-script document sample. To handle this, we propose a technique which is simple, rotation invariant, and robust to various complexities such as unconstrained gaps between words, lines, writing styles, skew angles and sizes.

Basic techniques are aimed to classify words, lines or text blocks from documents where a few different scripts are present, by using either global, local features or integrating both [1]. Global approaches are primarily based on DCT [12, 13], discrete wavelet transforms (DWT) and Gabor filters. Global approaches are efficient in representing large size images i.e., text blocks, for instance. They are faster, robust to noise and improper segmentation, and are script independent. For example, text blocks of eight Indic scripts are classified based on DCT and wavelet features [13]. In contrast, local approaches employ shape features based on connected components [4, 7, 8]. They are script dependent, slower in computation. These methods may not offer rotation invariant property and thus their usefulness can be limited. From practical point of view, rotation invariant classification of scripts is highly desirable [15] to enhance the performance of the OCR. On the whole, global approach is the better choice to deal with such problems by incorporating the appropriate modifications so that generic and optimized solutions [2] are possible. Under this purview, in this paper, we study directional DCT (D-DCT) to address the aforementioned challenges. We primarily aim to demonstrate how D-DCT [3], [16] is efficient while considering rotation invariant property through the statistics of directional energy distributions of DCT coefficients.

2 Materials and Method

In short, the proposed method (similar to [5]) can be described as follows. Words are first extracted from each document based on the morphological operators where we primarily are based on connected component. The extracted words are then represented with DCT features and its variants. For classification, a well-known LDA classifier is employed.

Feature selection. Before applying directional DCT, the primary task is to extract words from input document image. It is composed of three steps: (1) image binarization, (2) image dilation (both horizontal and vertical), and (3) word image extraction based on connected components. Further, the length of the structuring element is adoptive to the script of the document. The complete process of word extraction from Devanagari document is shown in Fig. 2.

We apply 2D DCT on each word image and compute its coefficient matrix as $C_{N \times N}$. Further, C is partitioned into three bands namely principal diagonal (μ), upper (α) and lower (β) diagonals of size $N - 2$. Then μ is extracted from C and computed standard deviation σ_1 using

$$\sigma_1 = \sqrt{\frac{1}{n-1} \sum_{u=1}^{n} C(u)_\mu - C(\bar{u})_\mu^2}, \tag{1}$$

where $u = 1, 2 \ldots\ldots n$ and n is the number of coefficients in μ. The standard deviation σ_1 is a scalar value. Similarly, α diagonals of C are extracted and computed their standard deviations using

$$\sigma_\alpha = \sqrt{\frac{1}{n-1} \sum_{u=1}^{n} C(u)_\alpha - C(\bar{u})_\alpha^2}, \tag{2}$$

where $u = 1, 2 \ldots\ldots n$ and $\alpha = 1, 2 \ldots N - 2$ and σ_α is column vector of size $N - 2 \times 1$. Then by appending the value of $\sigma 1$ and a zero into σ_α, we get first feature f_1 of dimension $N \times 1$. In the same way, β diagonals of C are extracted and computed their standard deviations using

$$\sigma_\beta = \sqrt{\frac{1}{n-1} \sum_{u=1}^{n} C(u)_\beta - C(\bar{u})_\beta^2}. \tag{3}$$

where $\beta = 1, 2 \ldots N - 2$ and σ_β is a column vector of size $N - 2 \times 1$. By appending two zeros into σ_β, we get second feature f_2 of dimension $N \times 1$. Similarly, features f_3 and f_4 are computed by flipping the input matrix C. The flipped matrix is denoted by C^f and upper, lower and principal diagonals are denoted by β^f, α^f and μ^f respectively. Finally, standard deviations of DCT of horizontal and vertical coefficients of C are computed to obtain features f_5 and f_6 respectively.

Thus, we have an integrated feature vector $F = \{f_{1 \ldots} f_6\}$ of size $N \times 6$. The dimension of the feature vector can further be reduced by taking their mean and

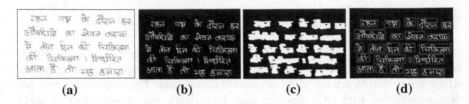

(a)   (b)   (c)   (d)

Fig. 2 An example showing word segmentation of Devanagari text block **a** Input Image, **b** Binarization, **c** Dilation, **d** Output

standard deviation i.e., the reduced dimension will be 12×1 from $N \times 6$ i.e., (six means and six standard deviations).

Classification. Since LDA offers class discriminating information to the higher extent by reducing dimensionality of feature space and also maximizes separability between the classes by maximizing the ratio of inter-class variance to the intra-class variance, we employ LDA and study its characteristics.

3 Experimental Results

Dataset and evaluation protocol. Our dataset is composed of 6,000 handwritten word images of six different scripts, namely Roman (Rom.), Devanagari (Dev.), Kannada (Kan.), Telugu (Tel.), Tamil (Tam.) and Malayalam (Mal.), 1,000 words of each script. Out of 6,000, 3,000 word images are reference text words (500 from each script), remaining 3,000 are rotated word images produced by choosing 100 word images of each script and rotated with various angles such as 30^0, 60^0... 150^0.

In order to evaluate the performance of the method, 10-fold cross validation has been implemented unlike traditional dichotomous classification. The performance of any script s classification is measured by using the precision,

$$Precision@s = \frac{number\ of\ correctly\ classified\ words}{total\ number\ of\ words}, s = \{1...6\} \qquad (4)$$

Results and discussions. To attest the performance of the proposed algorithm, tests are primarily carried out in three ways and provided in Tables 1 and 2. In Table 1, the first two issues are covered where results that are shown in

1. lower triangular part of the table are based on rotated word images i.e., 100 word images of each script and five different angles of each word image; and
2. upper triangular part of the table are based on non-rotated word images i.e., 500 word images of each script.

In Table 2, another issue i.e., the superiority of D-DCT over C-DCT will be confirmed by performing a test using non-rotated, rotated and mixed word images.

Table 1 Bi-script identification performance in % using D-DCT

Scripts	Eng.	Dev.	Kan.	Tel.	Tam.	Mal.	Avg.
Eng.	–	99.7	92.7	98.4	97.8	97.4	97.2
Dev.	93.8	–	99.5	99	98.9	99.7	99.28
Kan.	91.4	99.3	–	94.9	91.6	93.5	93.33
Tel.	97.9	99.6	93.7	–	97.9	98.2	98.05
Tam.	91.1	99.0	91.7	93.5	–	98.9	98.9
Mal.	98.5	99.7	93.8	98.3	96.3	–	97.35
Avg.	94.54	99.4	90.03	95.9	96.3	95.84	

Table 2 Performance comparison of D-DCT with C-DCT

Bi-Scripts	Non-rotated images		Rotated images		Mixed images	
	C-DCT	D-DCT	C-DCT	D-DCT	C-DCT	D-DCT
Rom./Others	86.30	97.20	70.38	94.54	67.80	92.98
Dev./Others	92.00	99.28	74.33	99.40	69.12	96.96
Kan./Others	80.00	93.33	71.90	90.03	71.17	88.90
Tel./Others	82.25	98.05	81.25	95.90	80.88	94.83
Tam./Others	84.00	98.90	78.25	96.30	76.25	96.70

The reported results in lower triangle of Table 1, highlights the significance of the directional energy distribution in classification of rotated word images. For instance 99.40 % average precision is achieved in classification of Devanagari versus other scripts. However, low average precision has been noticed in case of Roman versus other scripts and Kannada versus other scripts as 94.54 and 93.07 % respectively. This is due to the similarity in writing style and thus similar vertical energy distributions. The average precisions shown in upper triangular part of Table 1 show little higher performance compared to lower triangle precisions. This is because of the uneven distributions of DCT coefficients in case of rotated images. Hence its average precision is 1.51 % lesser than that. To compare the performance of D-DCT with C-DCT, tests are carried out on rotated, non-rotated and the combination of these word images and the results are provided in Table 2. The C-DCT yields average precision of 86.30 % with non-rotated word images of Roman script in combination with other five scripts whereas, D-DCT gives a superior result of 97.20 % for the same combinations. On the whole, the D-DCT yields 94.54 % in average in comparison to 70.38 % from C-DCT.

Comparative study. For comparison, we have also extended our experimentation on a dataset of 22,500 printed word images used in [12] and achieved the average identification accuracy of 97.06 %, which is higher in comparison to 93.5 % using LDA for multi-script identification. The results of [12] is achieved by using 36 features; however, we provide an accuracy of 97.06 % by using only one-third features of [12], i.e., 12 features. Besides, we have presented major state-of-the art methods to compare methods on a one to one basis in Table 3.

Table 3 Comparative analysis

Method	Major features	Scripts	Accuracy (%)
Sarkar et al. [14]	Horizontal, foreground and background features	Devanagari and Roman	99.28
		Bangla and Roman	98.43
Namboodiri and Jain [10]	Spatial and temporal features	Arabic, Cyrillic, Devanagari, Han, Hebrew and Roman	95.1
Hochberg et al. [6]	Relative X Centroid, holes, sphericity and aspect ratio	Arabic, Cyrillic, Devanagari, Chinese, Japanese and Roman	88.00
Moussa et al. [9]	Fractal features	Arabic and Latin	96.64
Our method	Directional DCT	Roman, Kannada, Telugu, Devanagari, Malayalam, Tamil	99.70

4 Conclusion

In this paper, we have studied the rotation invariant features based on directional DCT (D-DCT) for word-level handwritten script identification and validated with six major Indic scripts. The features used in this method are derived based on visual perception of the shape of characters of Indic scripts, which are basically dominated by directional strokes. One of the recent extensions is, we are working on printed word images of eleven Indic scripts used in [11] where the preliminary results are encouraging.

References

1. Belaid, A., Santhosh K.C., D' Andeey, V.P.: Handwritten and printed text separation in real document, In: Proceedings of Machine Vision Applications, pp. 218–221 (2013)
2. D Ghosh, T.D., Shivaprasad, A.P.: Script recognition a review. IEEE Trans. Pattern Anal. Mach. Intell. **32**(12), 2142–2161 (2010)
3. Fu, J., Zeng, B.: Directional discrete cosine transforms: A theoretical analysis. In: Proceedings of the IEEE International Conference on Acoustics, Speech, and Signal Processing, vol. 1, pp. 1105–1108 (2007)
4. Hangarge, M., Dhandra, B.V.: Offline handwritten script identification in document images. Int. J Comput. Appl. **4**(6), 1–5 (2008)
5. Hangarge, M.,Santhos, K.C., Rajmohan, P.: Directional Discrete Cosine Transform for Handwritten Script Identification. In: Proceedings of International Conference on Document Analysis and Recognition, pp. 344–348 (2013)
6. Hochberg, J., Bowers, K., Cannon, M., Kelly, P.: Script and language identification for handwritten document images. Int. J. Doc. Anal. Recogn. **2**(2–3), 45–52 (1999)
7. Roy, K., Banerjee, A., Pal, U.: Word-wise hand-written script separation for Indian postal automation. In: Proceedings of International Workshop on Frontiers in Handwriting Recognition, pp. 521–526, (2006)

8. Zhou, L., Lu, Y., Tan, C.L.: Bangla/english script identification based on analysis of connected component profiles. In: Proceedings of International Conference on Document Analysis and Recognition, pp. 243–254 (2006)
9. Moussa, S.B., Zahour, A., BenAbdelhafid, A., Alimi, A.M.: Fractal-based system for Arabic/ Latin, printed/handwritten script identification. In: Proceedings of International Conference on Pattern Recognition, pp.1–4 (2008)
10. Namboodiri, A., Jain, A.: Online handwritten script recognition. IEEE Trans. Pattern Anal. Mach. Intell. 26(1), 124–130 (2004)
11. Pati, P.B., Ramakrishnan, A.G.: Word level multi-script identification. Phys. Rev. Lett. 29(9), 122–1218 (2008)
12. Pati, P.B., Ramakrishnan, A.G.: Hvs inspired system for script identification in Indian multi-script documents. In: Proceedings of International Conference on Document Analysis and Recognition, pp. 380–389 (2006)
13. Rajput, G.G., Anitha, B.H.: Handwritten script recognition using DCT and wavelet features at block level. Int. J Comput. Appl. 158–163 (2010)
14. Sarkar, R., Das, N., Basu, S., Kundu, M., Nasipuri, M., Basu, D.K.: Word level script identification from Bangla and Devanagri handwritten texts mixed with Roman script. J Comput 2(2), 103–108 (2010)
15. Tan, T.: Rotation invariant texture features and their use in automatic script identification. IEEE Trans. Pattern Anal. Mach. Intell. 20(7), 751–758 (1998)
16. Zeng, B., Fu, J.: Directional discrete cosine transforms: a new framework for image coding. IEEE Trans. Circuits Syst. Video Technol. 18(3), 305–313 (2008)

Selection of Graph-Based Features for Character Recognition Using Similarity Based Feature Dependency and Rough Set Theory

Sunanda Das, Suvra jyoti Choudhury, Asit Kumar Das and Jaya Sil

Abstract Recently, large amount of data is populated almost in every field, analysis of which is a challenging task in data mining community. Feature based character recognition is a well-known field of research where numerous features are used without analyzing their importance resulting lengthy recognition process. Feature selection plays an important role in character recognition problem which has not been explored. In the paper, the characters are represented by graphs and features of the graphs form feature vectors. A novel feature selection method has been proposed using the concepts of feature dependency and rough set theory to select only the features which are important for character recognition. Initially, feature dependency is measured based on correlation coefficients and similarity among the features are evaluated using feature dependency based on which the features are ranked. Rough set theory based quick reduct generation algorithm is applied for selecting the important features using feature ranking. The method is applied on character data set as well as on various benchmark data set and the experimental result is compared with well-defined dimension reduction techniques that demonstrates the effectiveness of the method.

Keywords Character recognition · Feature dependency · Similarity measure · Feature selection · Rough set theory

S. Das (✉)
Neotia Institute of Technology, Management and Science, Jhinga, Diamond Harbour Road,
South 24 Pargana 743368, India
e-mail: Sunanda_srkr@yahoo.co.in

S. j. Choudhury
Department of Purabi Das School of Information Technology, Bengal Engineering
and Science University, Shibpur, Howrah 711103, India
e-mail: cshuvrajyoti@gmail.com

A. K. Das · J. Sil
Department of Computer Science and Technology, Bengal Engineering and Science
University, Shibpur, Howrah 711103, India

G. P. Biswas and S. Mukhopadhyay (eds.), *Recent Advances in Information Technology*,   57
Advances in Intelligent Systems and Computing 266, DOI: 10.1007/978-81-322-1856-2_7,
© Springer India 2014

1 Introduction

Generally, huge volume of data set generated in every moment is not equally important for decision making. Hence, dimension reduction takes an important role to select only relevant features by discarding the insignificant one from the data set. Many problems in machine learning involve high dimensional descriptions of input as a result much research has been carried out on dimensionality reduction [1]. However, existing work tends to destroy the underlying semantics of the features after reduction or require additional information about the given data set for thresholding. A technique that can reduce dimensionality using information contained within the data set and that preserves the meaning of the features is highly desirable. Rough set theory (RST) [2, 3] is used to discover data dependencies and to reduce the number of features present in a dataset without any additional information. Graph classification is an important data mining task for which many kernel methods [4] are already implemented. Recently, in various fields including bio-informatics, chemistry and so on, graph data are propagated which can be classified using the machine learning and data mining approach.

Here, characters are represented as graphs and several features are extracted from the graph. Correlations among the features are computed based on which feature dependency is measured. For each feature A_t a feature dependency $A_t \to A_i, A_j, \ldots, A_k$ is computed. Then, similarity among two features is measured using corresponding dependencies and average similarity of a feature with respect to all other features is computed and the features are ranked based on their decreasing similarity factors. Finally, quick reduct algorithm of RST [2, 3] is applied on the features to generate the reduct based on their rank. The overall work is described below:

Procedure: Graph_Based_Feature_Extraction
INPUT: Scanned image of a character.
OUTPUT: Reduced feature set or Reduct.
BEGIN:
Step 1: Input an image and convert it into grey level image.
Step 2: Apply canny edge detection to get edge map image.
Step 3: Contours or curves extraction from edge map image.
Step 4: Select control points from extracted curves.
Step 5: Generate graph using curves and control points.
Step 6: Extract features from the graph.
Step 7: Compute rank of the features based on similarity measurements.
Step 8: Select important features using "Quick Reduct Method" of Rough Set Theory.
Step 9: From reduct features particular character is recognized.
END.

The remaining part of this paper is organized as follows. Graph generation for characters and feature extraction from the graph is discussed in Sect. 2. Section 3 describes reduct generation based on feature dependency and RST. Experimental results are demonstrated in Sect. 4 and finally conclusion is made on Sect. 5.

2 Graph Based Feature Extraction

English characters both in lower and upper cases and numeral '0' to '9'written with different style are collected [5] and used for the work of character recognition. Each character is considered as an image and image processing concepts are applied on it that yields a graph from which some graph based features are extracted [4] for character.

2.1 Boundary Detection

Save the scanned copy of characters in an image file obtained from [5] and convert the image file into grey level image. Then canny edge (level) detection method is applied to obtain binary edge maps for the set of characters. The process is illustrated using the images, shown in Fig. 1.

The edge map image contains the boundary information of the image. To detect boundary from the edge map image, pixels with value '1' are searched along the boundary of the character and stored respective position of the points. When a point with intensity '1' is detected, traverse from that particular position considering a neighbor around the point and select the closest neighbor points and repeat the process. After searching all the neighbor points, the points containing '1' are stored in the image array which represents a contour of a character image extracted from the edge map image. The process continues to extract other contours of the image and finally terminates when no more points with intensity '1' is present in the edge map image. For example the process extracts three contours for 'B', two contours for 'a' and two contours for '9' as shown in Fig. 2.

2.2 Control Point Selection, Graph Generation and Feature Extraction

Now from these extracted curves select several control points to maintain the shape of the curve. Curvature [6] at a low scale for each contour is computed to retain all true corners, which are points on the curve with maximum local curvature. Then rounded corners [7] and false corners are eliminated and end points of

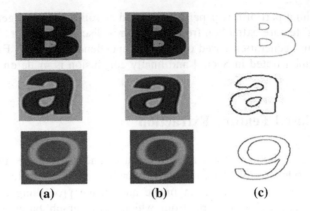

Fig. 1 **a** Original image; **b** grey level image; **c** edge map image

Fig. 2 Contours for images 'B' 'a' and '9'

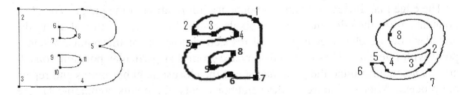

Fig. 3 Control point selection from contours of **a** 'B'; **b** 'a'; **c** '9'

line mode curve are selected as corners to remove the boundary noises, which gives all possible control points in the curve. The method selects control points for each contour of the image as shown by bold dot points in Fig. 3. Finally, starting from one control point traverse along the curve until another control point is found which gives an edge of the graph. Continuing the traversal process exhausting all the control points of a contour and gives a component of the graph. Finally when all control points corresponding to an image are exhausted, a graph is obtained. So the graph for a character image is connected if there is only one contour of the image; otherwise disconnected with components equal to its number of contours. Then for each graph extracts features and their values, as listed in Table 1.

Table 1 Extracted features and their values for three character images

Extracted features	Feature values for		
	Graph of 'B'	Graph of 'a'	Graph of '9'
Average degree	2	1.78	1.75
Average clustering coefficient	0.45	0.00	0.00
Average effective eccentricity	1	1	1
Percentage of end points	0	22.22	12.50
Number of nodes	11	9	8
Number of edges	11	8	7
Trace	13.47	12.99	12.99
Spectral radius	2	2	2
Energy	21	20	23
No of components	3	2	2
Label entropy	0	0	0
Neighborhood Impurity	0	0	0
Number of eigen values	11	10	10

3 Reduct Generation Based on Feature Dependency

Let $DS = (U, A)$ be a dataset, where U is the set of characters and $A = \{A_1, A_2, ..., A_n\}$ is the set of features. Now correlation $R_{A_iA_j}$ between two features A_i and A_j in A is computed using Eq. (1), where, $\bar{A}_i$, $\bar{A}_j$, are mean and σ_{A_i}, σ_{A_j}, the standard deviation of A_i and A_j respectively.

$$R_{A_iA_j} = \frac{\sum (A_i - \bar{A}_i)(A_j - \bar{A}_j)}{n\sigma_{A_i}\sigma_{A_j}} \tag{1}$$

Using Eq. (1) correlations between all pair of features are obtained and a correlation matrix is built. A feature dependency set FS is defined where elements in the set are of the form $\{A_i \rightarrow A_{j_1}A_{j_2}, ..., A_{j_k}\}$ where, A_i is the feature corresponding to ith row of the matrix and $A_{j_1}, A_{j_2}, ..., A_{j_k}$ are the features correlated to A_i. Let the set FS contains n feature dependencies. From consequent part of feature dependency $\{A_i \rightarrow A_{j_1}A_{j_2}, ..., A_{j_k}\}$, A_i is removed as $A_i \rightarrow A_i$ is the trivial dependency, for all $i = 1$ to n.

3.1 Importance of Features

All features are not equally important and so only important features are kept in the system. The importance of the features is measured using RST based quick reduct generation algorithm. Similarity among the features is measured using FS. For any feature dependency $A_1 \rightarrow B$ antecedent A_1 is a single feature and consequent B is a collection of features. Similarity factor S_{ij} between two features A_i

and A_j are computed using corresponding feature dependencies $A_i \rightarrow B$ and $A_j \rightarrow C$ by the relation $S_{ij} = (B \cap C)/(B \cap C)$. $S_{ij} = 0$ implies that A_i and A_j are not related to any common feature, hence considered as dissimilar feature and $Sij = 1$ implies that both are related to same set of features and so they are highly correlated. Obviously, $0 \leq S_{ij} \leq 1$ and greater its value implies higher the correlation among A_i and A_j. Higher average similarity factor for a feature implies that the feature is related to more number of features and so it is more important or higher ranked feature.

3.2 Quick Reduct Generation

Reduct [2, 3] is a sufficient set of features which by itself fully characterize the knowledge in the dataset. The information system projected on these features possesses the same equivalence class structure as that expressed by the full feature set. Let $DS = (U, A, C, D)$ be a decision system where U is the finite, non-empty set of objects or characters, A is a finite, non-empty set of features such that $A = C \cup D$ and $C \cap D = \emptyset$, C and D are set of condition and decision features, respectively. Attribute *dependencies* describes which features are strongly related to which other features which in RST is defined using the positive region as described by Eq. (2).

$$\gamma_P(D) = \frac{|POS_P(Q)|}{|U|} \tag{2}$$

Thus, a reduct R can be thought of as a sufficient set of features if $\gamma_R(D) = \gamma_C(D)$. In other words, R is a reduct if the dependency of decision feature D on R is exactly equal to that of D on whole conditional feature set C. The quick reduct algorithm proposed in [2, 3] starts with any arbitrary feature and continue the process until a reduct is found. But here, the features are arranged according to their rank obtained by Sect. 3.1. So the algorithm starts with the highest ranked feature as possible reduct and check if $\gamma_R(D) = \gamma_C(D)$, a reduct is obtained; otherwise next highest ranked feature is considered together with the possible reduct to check for reduct and so on. Since there may be many features with same rank, so many reducts may be formed.

4 Result and Discussions

The method is applied on character data set made using characters collected from [5], where, number of each of 62 types of characters is 70 with different fonts and orientations, total number of characters is 4340 and number of features is 13. The

Table 2 Illustrate the accuracies of various classifiers on reduced character data set

Classifier	Origina data set (13)	CFS (7)	CON (8)	Proposed method (7)
Native bayes	74.67	76.73	76.75	77.50
KSTAR	78.01	78.12	75.62	77.50
Bagging	73.53	75.62	75.62	77.50
Multi class classifier	76.73	76.25	77.50	78.12
J48	77.21	76.87	77.50	78.12
PART	77.19	76.87	77.50	77.50
Multilayer perceptron	74.98	75.65	76.87	78.75
Average accuracy (%)	76.05	76.59	76.77	77.86

method selects seven features {'Number of eigenvalues', 'Energy', 'Trace', 'Number of edges', 'Number of nodes', 'Average clustering coefficient', 'Average Degree'}. Number of features selected by the method and the accuracies of the classifiers based on reduced dataset are quite impressive. Table 2 lists the number of features and accuracy of the classifiers on original and reduced data set, where datasets are reduced by proposed as well as existing methods CON [8] and CFS [9]. The methods CFS and CON and the classifiers are run using WEKA tool [10].

5 Conclusion

The work represents characters into graph and extracts various features from the graph for character recognition. It is a novel concept of character recognition in the field of data mining and rough set theory. The result obtained is quiet impressive, but it can be improved by extracting some other features and also by increasing volume of the data set. As a future work, multi objective genetic algorithm will be applied for feature optimization and also the result obtained will be compared with some other existing character recognition techniques.

References

1. Dash, M., Liu, H.: Feature selection for classification. Intell. Data Anal. **1**(3), 131–156 (1997)
2. Pawlak, Z.: Rough Sets: Theoretical Aspects of Reasoning About Data. Kluwer Academic Publishing, Dordrecht (1991)
3. Polkowski, L.: Rough Sets: Mathematical Foundations. Advances in Soft Computing. Physica Verlag, Heidelberg, Germany. 2002
4. Li, Geng., Semerci[†], M., Yener, B., Zaki, M.J.: Graph classification via topological and label features
5. http://www.ee.surrey.ac.uk/CVSSP/demos/chars74k/
6. He, X.C., Yung, N.H.C.: Curvature scale space corner detector with adaptive threshold and dynamic region of support. In: Proceedings of the 17th international conference on pattern recognition, vol. 2, pp. 791–794, August 2004

7. He, X.C., Yung, N.H.C.: Corner detector based on global and local curvature properties. Opt. Eng. **47**(5), 057008 (2008)
8. Yu, L., Liu, H.: Efficient feature selection via analysis of relevance and redundancy. J. Mach. Learn. Res. **5**, 1205–1224 (2004)
9. Hall, M.A.: Correlation-based feature selection for machine learning. Dissertation, The University of Waikato (1999)
10. Hall, M., Frank, E., Holmes, G., Pfahringer, B., Reutemann, P., Witten, I.H.: The WEKA data mining software: an update. SIGKDD Explor. **11**(1), 10–18 (2009)

A Partial Image Cryptosystem Based on Discrete Cosine Transform and Arnold Transform

Kshiramani Naik and Arup Kumar Pal

Abstract Partial image cryptosystem has drawn much more attention in various real time applications in order to address issue like security along with the computational overhead. Generally in the partial image cryptosystem, the encryption-decryption process is employed on the compressed version of secret image (i.e. the significant part of the secret image) for reducing the computational overhead. In this paper, we have proposed a partial image cryptosystem for uncompressed color image where the discrete cosine transform is employed on each color components of the color image to select the significant coefficients and subsequently the selected coefficients of the color image are fed into encryption process for reducing the computational overhead. In encryption process, the selected coefficients are confused using Arnold transform followed by diffusion with keys. After completion of the encryption process, unencrypted coefficients are appended with encrypted components to form the uncompressed encrypted image. The proposed scheme has been tested on a set of standard color test images and satisfactory results have been found. In addition, the simulation results show the effectiveness of the proposed image cryptosystem in terms of security analysis.

Keywords Arnold transform · Confusion–diffusion · Discrete cosine transform · Partial image cryptosystem

K. Naik (✉) · A. K. Pal
Department of Computer Science and Engineering, Indian School of Mines,
Dhanbad, Jharkhand 826004, India
e-mail: kshiramani@gmail.com

A. K. Pal
e-mail: arupkrpal@gmail.com

G. P. Biswas and S. Mukhopadhyay (eds.), *Recent Advances in Information Technology*,
Advances in Intelligent Systems and Computing 266, DOI: 10.1007/978-81-322-1856-2_8,
© Springer India 2014

1 Introduction

The interest of usage of digital information has increased rapidly due to the advancement of Internet technology. Internet itself is not a secured communication channel, so the illegal accessing of data is a common threat. As a result, it is necessary to employ some security mechanism on the secret data before transmitting through the Internet. This digital information not only comprises text, but also has multimedia data like image, audio, and video, which are comparatively very bulky than the textual data [1]. In general, digital images have wide application in various fields, so it has become an important issue to provide the security on image data. The most effective method to protect the image data from being illegal access is done by employing suitable encryption techniques so that the only authorized entities with the key can decrypt them. Although several well accepted standard conventional data encryption techniques like DES, AES, RSA, ECC etc. have been used for protecting the textual data [2] but they are not suitable to employ directly on the image due to its bulky size and strong correlation among the adjacent pixels [3]. The general architecture of image encryption technique comprises of confusion process followed by diffusion process. The confusion cast the image elements into mix-up by changing the position of the pixels in such a way that the original image is not recognizable. The confusion is done by various reversible techniques based on Arnold cat transform (ACM) [4], magic square transform [5], chaos system [6] etc. In diffusion stage, the pixel values are changed by using various cryptographic algorithms like SCAN based methods [7], chaos based methods [8], tree structure based methods [9] etc. In general, the encryption-decryption process for image data is carried out either considering the whole data set i.e. known as total/full encryption-decryption method or by selecting significant elements of the secret image i.e. known as selective/partial encryption-decryption method [10, 11]. Selective/partial encryption has many advantages in real time applications [11]. First, it reduces the encrypted data volumes and improves the efficiency by reducing the computational overhead. Second, some format information can be left unchanged, which is compliant with compression or communication. Hence the partial encryption attracts much more attention in various real time applications to address the issues like security along with speed and efficiency. In general, the selective/partial image cryptosystem is carried out in two stages. In the first stage, the image data are compressed using some suitable transformation tools like DCT [12], DWT [4], PCA [13], SVD [14] etc. while in the subsequent stage, the compressed image data are encrypted using suitable encryption algorithms. Many researchers have already proposed various selective/partial image cryptosystems in the compressed domain in order to achieve effectiveness [14, 15]. But these encryption techniques are applicable only on lossy images and are not suitable on some application like medical sciences where the medical images are preserved as lossless manner. So in this paper, our intention is to design a partial image cryptosystem for uncompressed color images.

The rest of this paper is organized as follows. Section 2 discusses the proposed image encryption algorithm in detail. The simulation results and security analysis are presented in Sect. 3. Finally the conclusions are stated in Sect. 4.

2 Proposed Color Image Cryptosystem

The proposed color image cryptosystem is designed in three major phases like Pre-processing, Key matrix generation and Encryption respectively. The details of the methodology used in the proposed cryptosystem are presented as follow.

2.1 Pre-processing phase

In this phase, the significant and the insignificant part of each color components of the color image is classified. The fundamental steps are presented as follows.

Step 1: Decompose the input color image A of size $N \times N \times 3$ into three color components, denoted as R, G, B of size $N \times N$ respectively.
Step 2: Divide each color component into non-overlapping blocks of size $n \times n$.
Step 3: Employ 2D-DCT (given in Eq. 1) on each block.

$$F_k(u, v) = \frac{c(u)c(v)}{4} \sum_{i=0}^{7} \sum_{j=0}^{7} f_k(i,j) \cos\left[\frac{(2i+1)}{16} u\pi\right] \cos\left[\frac{(2j+1)}{16} v\pi\right] \quad (1)$$

$$C(e) = \begin{cases} \frac{1}{\sqrt{2}} & \text{if } e = 0 \\ 1 & \text{if } e \neq 0 \end{cases}$$

where $f_k(i, j)$ and $F_k(u, v)$ are the pixel value at position (i, j) and the DCT coefficient at coordinate (u, v) of block k respectively.
Step 4: Perform the zigzag scanning (as shown in Fig. 1) on each block to produce one dimensional vector of length n^2 in which the coefficients are arranged in decreasing order of energy.
Step 5: Divide each transformed vector as significant and insignificant vector. Select first t^2 number of coefficients from each transformed vector as significant and $n^2 - t^2$ number of coefficients as insignificant.
Step 6: Reshape significant coefficients into a square block of size $t \times t$.
Step 7: Combine the significant blocks in each color component to form a DCT-based significant matrix, of size $S \times S$ (where $S = \frac{Nt}{n}$) and denoted as R_S, G_S, B_S for Red, Green and Blue component respectively.
Step 8: The matrices R_S, G_S, B_S are kept as the form of unsigned integer of 8 bits.

Fig. 1 Zigzag scanning
order

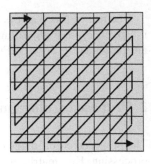

2.2 Key Matrix Generation Phase

In the proposed scheme, we need a key matrix of size S × S and the elements of the matrix will be unsigned integer of 8 bits. Here, we have taken a secret key, K of size 512 bits. This key is transformed into a key matrix of size S × S. The steps of the key matrix generation are given below.

Step 1: Convert the key, K into a block matrix of size 8 × 8 where the elements of the matrix is an unsigned integer of 8 bits.

Step 2: Concatenated the same block several times as a raster scan order to produce a large matrix, K_t of size T × T where T ≥ S. Truncate some rows and columns of the matrix, K_t so that it becomes the size of S × S.

Step 3: To remove the symmetric pattern of the matrix obtained in earlier step, the ACM (as given in Eq. 2) will be employed on the key matrix. The ACM will be applied three times with different parameters for generating three keys matrices i.e. K_R, K_G and K_B for encryption purposes of Red, Green and Blue components respectively.

2.3 Encryption Phase

Step 1: Confuse R_S, G_S, B_S by employing ACM as given in Eq. 2 to generate matrix R_S', G_S' and B_S' respectively. The parameters p, q and m (Number of iteration) are kept as secret keys.

$$\begin{bmatrix} x' \\ y' \end{bmatrix} = A \begin{bmatrix} x \\ y \end{bmatrix} \bmod N \tag{2}$$

where $A^{-1} = \begin{bmatrix} 1 & p \\ q & 1+pq \end{bmatrix}$, p, q are positive integers, $(x, y) \in [1, N]$ and (x, y), (x', y') represent the position vector of image pixel shifted before and after, respectively, and mod denotes the modulus after division.

Step 2: Encrypt the confused color components by employing XOR operation with their corresponding keys.

$$ER_S = R'_S \oplus K_R$$
$$EG_S = G'_S \oplus K_G$$
$$EB_S = B'_S \oplus K_B$$

Step 3: Append the insignificant matrices of each color component with individual encrypted R, G, B component to get R_E, G_E and B_E respectively.

Step 4: Apply inverse DCT (as given in Eq. 3) to each color component.

$$f_k(i,j) = \frac{1}{4} \sum_{u=0}^{7} \sum_{v=0}^{7} c(u)c(v)F_k(u,v) \cos\left[\frac{(2i+1)}{16}u\pi\right] \cos\left[\frac{(2j+1)}{16}v\pi\right] \quad (3)$$

Step 5: Combine the encrypted parts to get the cipher image A$_E$.

The schematic diagram of the proposed encryption process is given in Fig. 2.

2.4 Decryption Phase

In the decryption phase, the reverse process of the encryption phase is applied on the cipher image to obtain the original secret image. The cipher image and the secret parameters used in encryption phase are given as input in decryption algorithm. As it is a symmetric key cryptosystem, the encryption and decryption keys are same, and expressed as (key: p, q, m and K_R, K_G, K_B), where (p, q, m) are the ACM parameters and K_R, K_G, K_B are the generated key matrix for R, G and B component respectively. The detail of the decryption procedure is given as follows:

Step 1: Decompose the cipher image, A$_E$ into three individual color components, R$_E$, G$_E$ and B$_E$.

Step 2: Divide each color component into non-overlapping blocks of size n × n.

Step 3: Employ 2D-DCT (given in Eq. 1) on each block.

Step 4: Perform the zigzag scanning on each block to select significant coefficients and insignificant Coefficients.

Step 5: Reshape significant coefficients into a square block.

Step 6: Combine the significant blocks in each color component to form a DCT-based significant matrices ER_S, EG_S and EB_S.

Step 7: Apply the XOR operation on the significant matrices of the color components with their corresponding key matrices.

$$R_S' = ER_S \oplus K_R$$
$$G_S' = EG_S \oplus K_G$$
$$B_S' = EB_S \oplus K_B$$

Step 8: Apply inverse ACM (as given in Eq. 4) to the significant blocks.

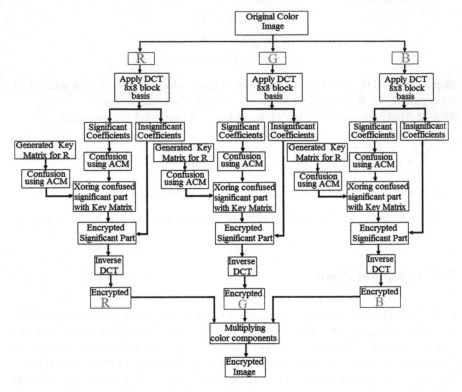

Fig. 2 Block diagram of encryption process

$$\begin{bmatrix} x \\ y \end{bmatrix} = A^{-1} \begin{bmatrix} x' \\ y' \end{bmatrix} \bmod N \qquad (4)$$

where $A^{-1} = \begin{bmatrix} pq+1 & p \\ -q & 1 \end{bmatrix}$

Step 9: Append the insignificant parts to their corresponding significant parts.
Step 10: Apply inverse DCT (as given in Eq. 3) to the significant blocks.
Step 11: Combine the three color components to get the original image A.

3 Experimental Analysis

This section presents the simulation results to evaluate the performance of the proposed cryptosystem. Analysis of the experiments is done on two standard color images i.e. Lena and Pepper of size $512 \times 512 \times 3$. The original and the

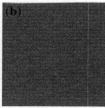

Fig. 3 Encrypted output for test images. **a** Original Lena, **b** Encrypted Lena, **c** Original pepper, **d** Encrypted pepper

Table 1 PSNR obtained for proposed cryptosystem

Image	PSNR
Lena	10.1530
Peppers	10.4751

corresponding encrypted images are shown in Fig. 3. As per the visual observation of human being, the encrypted images appeared as noisy image.

Also the degradation can be evaluated using PSNR. The PSNR for RGB color image is calculated as follows:

$$PSNR = 10 \log_{10}\left(\frac{255^2 \times 3}{MSE(R) + MSE(G) + MSE(B)}\right) db$$

$$MSE = \frac{1}{N^2} \sum_{r=0}^{N-1} \sum_{c=0}^{N-1} \left(f(r,c) - \tilde{f}(r,c)\right)^2$$

where f and $\tilde{f}$ denotes the original and reconstructed color component respectively, and the images are of size N × N.

As PSNR of 25 dB is the minimum threshold for perceptual similarity between any two images, the obtained PSNR values are compared with this value. The PSNR values are indicated in Table 1. It is observed that the PSNR obtained for the test images is less than 11 dB. This is well within the satisfactory limit. High perceptual degradation and low PSNR values obtained for the proposed technique reflects the satisfaction of the objective and subjective evaluation metrics.

To prevent the access of information to attackers, it is important to ensure that encrypted and original images do not have any statistical similarities. The histogram shows the statistical representation or distribution of pixel values within an image. The histogram of the cipher an image as shown in Fig. 4e–g, are significantly different from the respective histogram of the original image and hence it does not provide any clue to employ any statistical attack on the proposed image cryptosystem.

For an ordinary image having definite visual content, each pixel is highly correlated with its adjacent pixels either in horizontal, vertical or diagonal direction. However, an efficient image cryptosystem should produce the cipher image

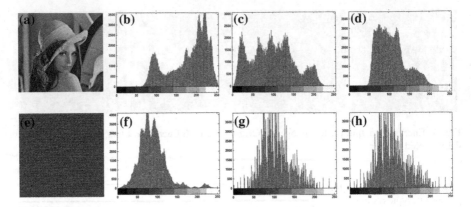

Fig. 4 Histogram of the color components. **a** Original image, **b** Histogram of original *red* component, **c** Histogram of original *green* component, **d** Histogram of original *blue* component, **e** Encrypted image, **f** Histogram of encrypted *red* component, **c** Histogram of encrypted *green* component, **d** Histogram of encrypted *blue* component

Table 2 Correlation analysis of RGB components of the color image

Direction	R		G		B	
	Original image	Cipher image	Original image	Cipher image	Original image	Cipher image
Horizontal	0.9798	0.4383	0.9691	0.5659	0.9327	0.5658
Vertical	0.9893	0.6516	0.9825	0.6048	0.9576	0.6048
Diagonal	0.9697	0.1810	0.9555	0.3857	0.9183	0.3857

with sufficiently low correlation in the adjacent pixels. From the Table 2, it is shown that the correlation value among the adjacent pixels of the encrypted images are comparatively low than the correlation value of adjacent pixels of original images.

In addition, the effectiveness of the proposed scheme is further analyzed based on the entropy calculation.

The entropy, H(m) of a message m can be measured by

$$H(m) = \sum_{i=0}^{M-1} p(m_i) \log_2 \frac{1}{p(m_i)},$$

where M is the total number of symbols $mi \in m$ and $p(m_i)$ is the probability of occurrence of symbol m_i.

For a random source emitting 256 equiprobable symbols, one has $H(m) = 8$ bits. Form Table 3, it is observed that the ciphered-images are close to random source and the proposed algorithm is secure against the entropy attack.

Table 3 Information entropy obtained for proposed cryptosystem

Image	Entropy
Lena	7.1062
Peppers	7.0289

4 Conclusion

This paper presents a partial color image encryption algorithm in DCT domain that avoids computational overhead during encryption and decryption process. The scheme is suitable to encrypt any uncompressed color images. Also this scheme will be suitable for enciphering medical images. In this work, we have also proposed a key matrix formation approach that is capable to form a large size of key matrix from a small size of binary key vector. The obtained encrypted images are completely different from the original images and even from performance analysis in terms of Histogram, PSNR, correlation and entropy, it is clear that the proposed scheme is capable to resist some cryptography attacks.

References

1. Lian, S.: Multimedia Content Encryption: Techniques and Applications, Taylor and Francis Group, LLC, Boca Raton (2009)
2. Schneier, B.: Cryptography: Theory and Practice, CRC Press, Boca Raton (1995)
3. Uhl, A., Pommer, A.: Image and Video Encryption: From Digital Rights Management to Secured Personal Communication, Springer, Ney York (2004)
4. Tanejaa, N., Ramanb, B., Guptaa I.: Selective image encryption in fractional wavelet domain, Int. J. Electron. Commun. (AEÜ) **65**, 338–344 (2011)
5. Lin, K.T.: Hybrid encoding method by assembling the magic-matrix scrambling method and the binary encoding method in image hiding, Optics Commun. **284**, 1778–1784
6. Kumar, A., Ghose, M.K.: Extended substitution–diffusion based image cipher using chaotic standard map, Commun. Nonlinear Sci. Numer. Simulat. **16**, 372–382 (2011)
7. Panduranga, H.T., Naveen Kumar, S.K.: A first approach on an RGB image encryption, Intern. J. Comput. Sci. Eng. **02**, 297–300 (2010)
8. Wang, X., Teng, Lin, Qin, Xue: A novel colour image encryption algorithm based on chaos. Sig. Process. **92**, 1101–1108 (2012)
9. Li, X., Knipe, J., Cheng, H.: A image compression and encryption using tree structures, Pattern Recognit. Lett. **18**, 1253–1259 (1997)
10. Furht, B., Kirovski, D. (eds.): Multimedia Encryption and Authentication Techniques and Applications. Auerbach Publications, Boca Raton (2006)
11. Lian, S., Chen, X.: On the design of partial encryption scheme for multimedia content. Math. Comput. Modell. **57**, 2613–2624 (2013)
12. Chang, C.C., Lin, C.C., Tseng, C.S., Tai, W.L.: Reversible hiding in DCT-based compressed images. Inf. Sci. **177**, 2768–2786 (2007)
13. Smith, L.: Atutorial on principal components analysis
14. Bhatnagar, G., Jonathan Wu, Q.M.: Selective image encryption based on pixels of interest and singular value decomposition. Digit. Signal Proc. **22**, 648–663 (2012)
15. Bahrami, S., Naderi, M.: Encryption of multimedia content in partial encryption scheme of DCT transform coefficients using a lightweight stream algorithm. Optik **124**, 3693–3700 (2013)

Part III
Technical Session-III: Topics in Algorithm Design

Swap Edges of Shortest Path Tree in Parallel

Anjeneya Swami Kare and Sanjeev Saxena

Abstract Let $G = (V, E)$ be a biconnected (2-edge connected), undirected graph with n vertices and m edges. A positive real weight is associated with every edge of the graph. Let d be the average depth, of a shortest path tree $S_G(s)$, rooted at s. Removal of a tree edge $e = (u, v)$ (u is parent of v) breaks the shortest path tree into two parts, T_1—the subtree containing s and T_2—the sub tree rooted at v. For each tree edge e, we are required to find a non-tree edge, with one end point in T_1 and the other end point in T_2 such that the average distance from the root s to all the nodes in the disconnected subtree T_2 is minimised. The proposed parallel algorithm can be implemented on the Concurrent Read Exclusive Write (CREW) model either in (1) $O((\log m)^{3/2})$ time using $O(m + nd + m(\log m)^{1/2})$ operations (processor-time product), or alternatively in (2) $O(\log m)$ time using $O(m + nd + m \log m)$ operations.

Keywords Swap edges · Replacement paths · Alternate paths

1 Introduction

Let $G = (V, E)$ be an undirected, biconnected (2-edge connected) graph, where V is the set of nodes and $E \subseteq V \times V$ is the set of edges. A positive real weight $w(e)$ is associated with each edge $e \in E$. Shortest path tree rooted at s will be

A. S. Kare (✉)
School of Computer and Information Sciences, University of Hyderabad,
Gachibowli Hyderabad 500046 Andhra Pradesh, India
e-mail: askcs@uohyd.ac.in

S. Saxena
Department of Computer Science and Engineering, Indian Institute of Technology,
Kanpur 208016 Uttar Pradesh, India
e-mail: ssax@cse.iitk.ac.in

G. P. Biswas and S. Mukhopadhyay (eds.), *Recent Advances in Information Technology*, 77
Advances in Intelligent Systems and Computing 266, DOI: 10.1007/978-81-322-1856-2_9,
© Springer India 2014

denoted by $S_G(s)$. Let S_v be the subtree of $S_G(s)$ rooted at v. The number of nodes in S_v will be denoted by $|S_v|$. The distance between any pair of nodes x and y in $S_G(s)$ will be denoted by $d(x, y)$; the distance between the nodes x and y after swapping the tree edge e with the non-tree edge f in $S_G(s)$ will be denoted by $d_{e/f}(x, y)$.

When edge $e = (u, v) \in S_G(s)$ (u is parent of v) is removed, then, the shortest path tree will be disconnected into two parts, T_1—the subtree containing s and T_2—the subtree rooted at v. The non-tree edges having one end point in T_1 and other end point in T_2 are called *crossing edges* for (the tree edge) e and will be denoted by C_e.

$$C_e = \{(x, y) \in E \backslash e \mid x \in T_1 \wedge y \in T_2\} \tag{1}$$

In order to restore the communication from s to all the nodes in S_v, we must use one of the crossing edges, say f, which we call a *swap edge*. As there can be as many as $O(m)$ possible candidate swap edges for a tree edge e, we must use a criteria to select one swap edge. In this paper, we consider the problem of minimising the average distance from the root to the nodes in the sub tree S_v [1].

Let $F(e,f) = \sum_{x \in S_v} d_{e/f}(s, x)$ be the sum of the distances of the nodes in S_v from s after swapping edges e and f. The average distance between s and the nodes in S_v when the tree edge e is swapped with the non-tree edge f is $\frac{F(e,f)}{|S_v|}$. Since $|S_v|$ is unchanged for a fixed v, the problem of finding the swap edge minimising the average distance from the root is equivalent to the problem of finding the swap edge minimising sum of the distances from the root.

The *best swap edge* f^* for the tree edge e is defined as the non-tree edge $f^* \in C_e$ for which, $F(e,f^*) \leq F(e,f) \forall f \in C_e$.

This paper is organised as follows: In Sect. 2 we discuss the sequential algorithm for the problem of computing swap edges minimising the average distance from the root. In Sect. 3 we discuss our parallel algorithm for the problem. We conclude with Sect. 4.

Nardelli et al. [1] described an $O(n^2)$ time algorithm for computing a swap edge for every tree edge that minimises average distance from the root s. Nardelli et al. also considered the problem of computing the swap edges minimising the distance between two pair of nodes. This problem is also known as Replacement Shortest Path (RSP) problem. For RSP, Nardelli et al. [2] described an $O(m \alpha(m, n))$ time algorithm. Mahadeokar and Saxena [3] described a linear time algorithm for integer weighted graph with small diameter. Earlier Malik et al. [4] described an $O(m + n \log n)$ time algorithm.

We describe a sequential algorithm to compute swap edges minimizing the average distance from the root. Our sequential algorithm is a modified version of the algorithm proposed in [1]. Complexity of our sequential algorithm is $O(m + nd) = O(n^2)$ same as that of algorithm of Nardelli et al., where d is the average depth of the shortest path tree.

The parallel implementation of our sequential algorithm takes $O((\log m)^{3/2})$ time using $O(m + nd + m(\log m)^{1/2})$ CREW operations or alternatively $O(\log m)$ time using $O(m + nd + m \log m)$ CREW operations.

2 Sequential Algorithm

The primary modification we make to the algorithm of Nardelli et al. [1] is in the computation of $F(e,f)$, so that it is more amenable to parallelisation. For the sake of completeness, we discuss the complete modified sequential algorithm. We use notations similar to the one used in [1].

2.1 Preliminary Computations

Let $e = (u, v)$ (u is parent of v) be a tree edge and let y be any descendant of v in $S_G(s)$. Apart from $|S_v|$, the following preliminaries are needed for the sequential algorithm:

count(v): will denote total number of nodes in S_u excluding the nodes in the subtree S_v.

$$count\,(v) = |S_u| - |S_v|. \tag{2}$$

Let l be the depth of the node v in $S_G(s)$. If $w_0 = s, w_1,...,w_l = v$ are the nodes in $S_G(s)$ on the path from the source s to the node v (see Fig. 1), then

$$|S_{w_i}| = \left(\sum_{j=l}^{i+1} count\,(w_j) + |S_v| \right) \tag{3}$$

down(v): will denote the sum of distances from v to every node in S_v.

$$down\,(v) = \sum_{y \in S_v} d(v, y) \tag{4}$$

up(v): will denote sum of the distances from u to the nodes in S_u excluding the distances of the nodes in the subtree S_v.

$$up\,(v) = \sum_{x \in S_u} d(u, x) - \sum_{x \in S_v} d(u, x) \tag{5}$$

minpath(y, v): is defined for every descendant y of v as (here, u is parent of v in $S_G(s)$).

$$minpath\,(y, v) = \min_{u'}\{d(s, u') + w(u', y)|(u', y) \in C_e(u, v)\} \tag{6}$$

Fig. 1 Ancestors of v and
non-tree edges incident at v

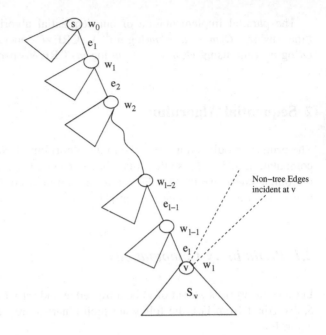

2.2 Computing $|S_v|$, $down(v)$, $up(v)$ and $count(v)$

Computation of $|S_v|$, $down(v)$, $up(v)$, $count(v)$ and $minpath(y, v)$ is done in a manner similar to [1].

Consider a node v in the post order traversal of the shortest path tree rooted at s. If v is a leaf, then $|S_v| = 1$, $down(v) = 0$. If v is not a leaf, let $v_1, v_2, v_3, \ldots, v_i$ be the children of v then

$$|S_v| = \sum_{j=1}^{i} |S_{v_j}| + 1 \tag{7}$$

$$down(v) = \sum_{j=1}^{i} \left(down(v_j) + |S_{v_j}| w(v, v_j) \right) \tag{8}$$

where summation is over each child v_j of v

$$up(v_j) = down(v) - down(v_j) - w(v_j, v)|S_{v_j}| \tag{9}$$

$$count(v_j) = |S_v| - |S_{v_j}|. \tag{10}$$

2.3 Finding Minpath Values

Consider a node v in the post order traversal of $S_G(s)$. Let $w_0 = s, w_1,...,w_l = v$ be the nodes in $S_G(s)$ on the path from the source s to the node v (see Fig. 1). Now consider the non-tree edges incident at v. For each such non-tree edge (x, v), compute $w_i = LCA(x, v)$, (for some $i, 0 \leq i \leq l$), where $LCA(x, v)$ is the Lowest Common Ancestor of x and v. The lowest common ancestor of any two nodes can be obtained in $O(1)$ time after linear time preprocessing [5].

Place this non-tree edge (x, v) in the bucket $B(v, w_i)$ with the "cost" or "key" as $\{d(s,x) + w(x,v)\}$. To find $minpath(v, w_j)$, $\forall \, 1 \leq j \leq l$, we find the minimum cost non-tree edge in the bucket $B(v, w_0)$, this is $minpath(v, w_1)$. Let $mincost[B(v, w_i)]$ be the cost of the minimum cost non-tree edge in the bucket $B(v, w_i), \forall i$.

Then for $2 \leq j \leq l$,

$$minpath\,(v, w_j) = \min\{minpath\,(v, w_{j-1}), mincost\,[B(v, w_j)]\} \qquad (11)$$

Each non-tree edge (x, y) $((y, x)$ is treated different from $(x, y))$ is labelled with a 4-tuple:

$$(x, depth(LCA(x,y)), LCA(x,y), \{d(s,y) + w(y,x)\})$$

and similarly for (y, x)

$$(y, depth(LCA(x,y)), LCA(x,y), \{d(s,x) + w(x,y)\})$$

There will be a total of $2(m - n + 1) = O(m)$ non-tree edges. Now we sort these non-tree edges, lexicographically, based on the first three parameters using bucket sort. As we know $S_G(s)$, $depth$ and LCA values, finding the label of a non-tree edge can be done in constant time. Thus all the non-tree edges can be labelled in $O(m)$ time. Sorting the non-tree edges based on the first three parameters of their label can again be done in $O(m)$ time using bucket sort. After sorting, all elements in a bucket will be together. Since the number of elements in all the buckets is bounded by $O(m)$, minimum cost non-tree edge in each and every (non-empty) bucket can be obtained in $O(m)$ time.

Using this sorted data and value of minimum cost in each of the buckets, the *minpath* values with respect to a fixed node v can be obtained in $O(l)$ time (if v has l ancestors for each of which *minpath* is defined). Finding all the *minpath* values will take $\sum l = nd$ where d is the average depth of the shortest path tree. Therefore the total time complexity for computing all the *minpath* values is $O(m + nd) = O(n^2)$.

2.4 Computing F(e, f)

Let v be any node in the post-order traversal of $S_G(s)$. Let $w_0 = s, w_1,...,w_l = v$ be the nodes in the path from s to v (see 1). Let $e_i = (w_{i-1}, w_i), 1 \leq i \leq l$, and let f_{il}

$_v = (x, v)$ be the crossing edge for e_i incident at v determining $minpath(v, w_i)$. The edge $f_{i/v}$ will be a swap candidate for the tree edge e_i.

To get $F(e_i, f_{i/v})$ we have to add distances from v to each node in S_{w_i}. As $minpath(v, w_i)$ is common in the computation of the length (from the root) to each node in S_{w_i}, $minpath(v, w_i)$ has to be added $|S_{w_i}|$ times in $F(e_i, f_{i/v})$. As $down(v)$ gives the sum of the distances of the nodes in S_v from v, then for the nodes in $S_{w_{j-1}} - S_{w_j}$ for $i + 1 \leq j \leq l$, the sum of $up(w_j)$ and $d(v, w_{j-1}) \times count(w_j)$ will give the sum of the lengths from v to each node in $S_{w_{j-1}} - S_{w_j}$. Thus, we proceed as follows:

1. $F(e_l, f_{l/v}) = down(v) + minpath(v, w_l)|S_v|$
2. For $r + 1 \leq i \leq l - 1$ (where r is such that $w_r = LCA(x, v)$)

$$F(e_i, f_{i/v}) = \sum_{j=l}^{i+1} \left(up(w_j) + count(w_j)d(v, w_j) \right)$$

$$+ \left(|S_v| + \sum_{j=l}^{i+1} count(w_j) \right) minpath(v, w_i) + down(v)$$

Next we look at an algorithm for computing $F(e_i, f_{i/v})$ where $e_i = (w_{i-1}, w_i), r + 1 \leq i \leq l$ for a fixed vertex v (see Algorithm 1). Here $f_{i/v} = (x, v)$.

Algorithm 1 Computing $F(e_i, f_{i/v})$ for fixed $v \neq s$

1: Let $w_0 = s, w_1, \ldots, w_l = v$ be the nodes in the path from s to v (see Fig. 1)
   Let w_r be the $LCA(x, v)$, where $f_{i/v} = (x, v)$
2: temp $= down(v)$, nodes $= |S_v|$
3: $F(e_l, f_{l/v}) =$ temp $+ minpath(v, w_l) \times$ nodes
4: **for** $i = l - 1$ to $r + 1$ **do**
5:    temp $=$ temp $+ up(w_{i+1}) + d(v, w_i) \times count(w_{i+1})$
6:    nodes $=$ nodes $+ count(w_{i+1})$
7:    $F(e_i, f_{i/v}) =$ temp $+ minpath(v, w_i) \times$ nodes
8: **end for**

Since $down(v)$, $|S_v|$, $minpath(u, v)$, $up(v)$ and $count(v)$ values have already been computed, each of these values can be accessed in $O(1)$ time. Therefore the algorithm will take $O(l)$ time for a fixed v (where l is the depth of the node v). Thus, for all n nodes the total time complexity will be $O(nd)$, where d is the average depth of the shortest path tree.

2.5 Computing Swap Edges

For each tree edge $(v, p(v))$ finding the non-tree edge which minimises sum of the distances from s to each node in S_v will take $O(|S_v|)$ time. This is because for each node w in S_v there can be at most one crossing (non-tree) edge for (tree edge) $(v, p(v))$ incident at w which determines the value $minpath(w, v)$.

The time complexity of the entire algorithm is $O(m + nd) + O\left(\sum_v |S_v|\right)$. Since each node v in the second summation will be counted l times, one for each of its l ancestors, hence $\sum_v |S_v| = \sum l = nd$, where d is the average depth of the shortest paths tree. Therefore, the time complexity of the algorithm is $O(m + nd)$.

Computing $|S_v|$, $down(v)$, $up(v)$ and $count(v)$ values takes $O(n)$ time. Computing $minpath$ values will take $O(m + nd)$ time. And computing swap edges for all the tree edges takes $O(m + nd)$ time. Thus the total time is $O(m + nd) = O(n^2)$.

3 Parallel Algorithm

We next consider the problem of parallelising the algorithm on the Concurrent Read Exclusive Write (CREW) model [6]. In this model, more than one processor can read from the same memory cell simultaneously, but no two processors can try to write in the same cell concurrently.

To parallelise the algorithm, we use the fact that the lowest common ancestor of any two nodes can be obtained in $O(1)$ time using a single processor after $O(\log n)$ time preprocessing using $n/\log n$ CREW processors [5].

Using the Euler tour technique pre-order numbering of a tree, depth of a tree and number of descendants of each node in a tree of n nodes can (all) be obtained in $O(\log n)$ time using $n/\log n$ CREW processors [5, 7].

Thus we get the values $|S_v|$. Values $down(v)$ can also be similarly obtained. Basically, we give a weight of 0 to edge $(p(v), v)$ and a weight of $w(v, p(v))|S_v|$ to edge $(v, p(v))$ in the Euler tour and carry out list ranking [8] followed by prefix sums computation [9] on this list, then

$$down(v) = \text{prefixsums}(v, p(v)) - \text{prefixsums}(p(v), v) \qquad (12)$$

Having $|S_v|$ and $down(v)$ values for all v, $up(v)$ and $count(v)$ can be obtained in constant time using one processors for each node v in the shortest path tree.

$$up(v) = down(p(v)) - down(v) - w(p(v), v)|S_v| \text{ and} \qquad (13)$$

$$count(v) = \left|S_{p(v)}\right| - |S_v| \qquad (14)$$

Thus, each non-tree edge (x, y) can be labelled with a 4-tuple:

$$(x, depth(LCA(x, y)), LCA(x, y), \{d(s, y) + w(y, x)\})$$

in constant time with a single processor (say associated with each of the non-tree edges).

Now sorting of the non-tree edges based on the first three parameters of their label can be done either in $O((\log m)^{3/2})$ time using $O(m(\log m)^{1/2})$ operations on CREW PRAM using the algorithm for integer sorting [10] or alternatively in $O(\log m)$ time with $O(m)$ processors using (general) parallel merge sorting procedure [11].

After sorting, all non-tree edges of the same bucket will be together. Finding $minpath(v, w_j)$ requires finding the minimum cost non-tree edge in each bucket $B(v, w_i)$ for $0 \leq i < j$. This can be obtained in constant time using range minima queries (after $O(\log m)$ time preprocessing using $O(m)$ CREW operations [12]).

Let $w_0 = s$, $w_1, \ldots, w_l = v$ be the nodes in the path from s to v and let $e_i = (w_{i-1}, w_i)$. Having all the minpath and $|S_v|$ values, each value in the set

$$\{minpath(v, w_l)|S_{w_l}|, \ minpath(v, w_{l-1})|S_{w_{l-1}}|, \ldots \ldots, minpath(v, w_1)|S_{w_1}|\}$$

is available in constant time. Consider the following two sets of values

$$\{0, \ d(v, w_{l-1})count(w_l), \ d(v, w_{l-2})count(w_{l-1}), \ldots \ldots, d(v, w_1)count(w_2)\}$$

and

$$\{down(v), \ up(w_l), \ up(w_{l-1}), \ldots, up(w_1)\}$$

Each individual element of either set is available in constant time. Both sets contains l elements, where l is the depth of the node v. The prefix sums on these sets can be carried out in $O(\log l)$ time using $l/\log l$ CREW processors [9]. Once the prefix sums are available, we can in $O(1)$ time compute:

$$F(e_i, f_{i/v}) = minpath(v, w_i)|S_{w_i}| + \text{prefixsums}(d(v, w_{i+1}) \times count(w_{i+1})) \\ + \text{prefixsums}(up(w_{i+1})) \tag{15}$$

It is clear that $f_{i/v}$ is a swap candidate for the tree edge e_i. Now as for every tree edge $(v, p(v))$ there are at most $|S_v|$ such candidates, an array of length $|S_v|$ for each tree edge $(v, p(v))$ is sufficient.

As $F(e_i, f_{i/v})$ is a candidate for the tree edge e_i, we store $F(e_i, f_{i/v})$ in the array associated with the tree edge e_i at index $(pre(w_i) + |S_{w_i}| - pre(v) + 1)$, where $pre(v)$ is the pre-order number of v. Therefore computing $F(e_i, f_{i/v})$ values for all the nodes can be done in $O(\log n)$ time using $\sum(l/\log l)$ CREW processors (as l is bounded by n). After computing $F(e_i, f_{i/v})$ values for all the nodes, the non-tree edge associated with the minimum values in the array for e_i will be a swap edge for e_i. These minimum values can be obtained in $O(\log (|S_v|))$ time using $|S_v|/\log|S_v|$ CREW processors [9]. Therefore computing swap edge for every tree edge can be done in $O(\log n)$ time using $\sum(|S_v|/\log|S_v|)$ CREW processors.

4 Conclusions

In this paper we considered the problem of finding the swap edge which minimises the average distance from the root to all the nodes in a shortest path tree of a given graph.

The sequential algorithm takes $O(m + nd)$ time (where d is the average depth of the shortest path tree).

The algorithm can be parallelised to get a parallel algorithm which takes either $O((\log m)^{3/2})$ time using $O(m + nd + m(\log m)^{1/2})$ CREW operations or alternatively $O(\log m)$ time using $O(m + nd + m \log m)$ CREW operations.

To our knowledge this is the first time a parallel algorithm has been proposed for this problem.

References

1. Nardelli, E., Proietti, G., Widmayer, P.: Swapping a failing edge of a single source shortest paths tree is good and fast. Algorithmica **35**(1), 56–74 (2003)
2. Nardelli, E., Proietti, G., Widmayer, P.: A faster computation of the most vital edge of a shortest path. Inf. Process. Lett. **79**(2), 81–85 (2001)
3. Mahadeokar, J., Saxena, S.: Faster replacement paths algorithms in case of edge or node failure for undirected, positive integer weighted graphs. J. Discrete Algorithms **23**, 54–62 (2013)
4. Malik, K., Mittal, A.K., Gupta, S.K.: The K most vital arcs in the shortest path problem. Oper. Res. Letters **8**, 223–227 (1989)
5. Schieber, B., Vishkin, U.: On finding lowest common ancestors: simplification and parallelization. SIAM J. Comput. **17**(6), 1253–1262 (1988)
6. JaJa, J.: Introduction to Parallel Algorithms. Addison-Wesley, Boston (1992)
7. Das, S.K., Chen, C.Y.: Cost-optimal parallel algorithms for traversing trees. In: IEEE Proceedings of SOUTHEASTCON '91 (Cat. No.91CH2998-3), vol. 1, pp. 474–478, (1991)
8. Anderson, R.J., Miller, G.L.: Deterministic parallel list ranking. Algorithmica **6**(6), 859–868 (1991)
9. Cole, R., Vishkin, U.: Deterministic coin tossing with applications to optimal parallel list ranking. Inf. Control **70**(1), 32–53 (1986)
10. Albers, S., Hagerup, T.: Improved parallel integer sorting without concurrent writing. Inf. Comput. **136**(1), 25–51 (1997)
11. Cole, R.: Parallel merge sort. SIAM J. Comput. **17**(4), 770–785 (1988)
12. Berkman, O., Schieber, B., Vishkin, U.: Optimal doubly logarithmic parallel algorithms based on finding all nearest smaller values. J. Algorithms **14**(3), 344–370 (1993)

A New Approach for Parallel Discrete Fourier Transform in Multi Mesh Network

Amit Datta and Mallika De

Abstract A simple parallel algorithm is proposed in this paper to compute Discrete Fourier Transform (DFT) coefficients of N points on Multi-Mesh (MM) architecture having N^2 processors where each of the processor generates a single element of the Fourier matrix based on the node position in the network i.e. Fourier elements are generated in place. The algorithm given here transforms a vector of length N in constant time multiplication and O $(\sqrt{N})$ addition and data communication time.

Keywords Multi mesh · Discrete Fourier transform · Parallel discrete Fourier transform · 2D mesh

1 Introduction

Discrete Fourier transforms (DFTs) [1, 2] are used extensively in a range of computing applications like audio processing, image processing, medical imaging etc.

A number of parallel algorithms developed for DFT/FFT in different architectures can be found in the literature [2–7].

Using an extension of the common factor algorithm authors in [3] have shown that a simple planar 2-dimensional systolic array having $N = M^2$ processing

A. Datta (✉)
Research Scholar, Department of Engineering and Technological Studies,
University of Kalyani, Kalyani, India
e-mail: amitdatta_wb@yahoo.co.in

M. De
Department of Engineering and Technological Studies, University of Kalyani, Kalyani,
India
e-mail: demallika@yahoo.com

G. P. Biswas and S. Mukhopadhyay (eds.), *Recent Advances in Information Technology*,   87
Advances in Intelligent Systems and Computing 266, DOI: 10.1007/978-81-322-1856-2_10,
© Springer India 2014

elements can be used to compute DFT of size N in $2M + 1$ steps of pipelined operations, achieving the area-time complexity $AT^2 = O (N^2 \log^3 N)$.

In this paper we present an entirely different but simple direct matrix–vector multiplication approach for computing DFT in Multi-Mesh architecture. The time complexity of the algorithm is $O (\sqrt{N})$ for N point DFT. The algorithm is implemented in Multi-Mesh (MM) architecture [7–9] having N^2 processors where each processor is of 4 degree. As shown in Fig. 1, there are N meshes of size $\sqrt{N} \times \sqrt{N}$ each, which themselves are again arranged in $\sqrt{N}$ rows and $\sqrt{N}$ columns so that there will be N^2 processors in the MM network.

The important aspect of the proposed parallel algorithm is that it requires constant time multiplication. The order of time complexity is mainly governed by the data communication and addition times. The Fourier components are generated in place in parallel for each processor using the processor position co-ordinates and require three multiplication steps and two addition steps. The input of the vector to be transformed and their propagation requires $(2n + 1)$ communication steps. The partial products are generated by a single multiplication step. After which $2(n - 1)$ additions and $(n + 1)$ data communication steps are required to get the transformed vector.

2 Discrete Fourier Transform

The DFT [1, 2] of a continuous function $a (t)$ is given by

$$A_f = \sum_0^{N-1} a_t e^{\frac{2\pi i f k}{N}}, \ 0 \leq f \leq N - 1$$

where, $i = \sqrt{-1}$. The variable t is used to represent time, and f is used to represent frequency. The Fourier transform is used to convert a function of time into a function of frequency.

3 The Multi Mesh (MM) Network

First, the MM network proposed by the authors [7] is described. The basic building block of the MM network is n × n mesh.

A processor inside a given block can be uniquely identified by two coordinates. Again blocks are organized as matrix form so each block can be identified by two coordinates, say α and β as M (α, β). Thus, each of the n^4 processors in MM can be uniquely identified using a 4-tuple. If the boundary processors of P (α, β_1) are connected to boundary processors of P (α, β_2) for all values of α, $0 \leq \alpha \leq n - 1$, we denote these sets of links by an interconnection between the sets P $(*, \beta_1)$ and P $(*, \beta_2)$. Inter-block vertical connection is identified by following rule.

Fig. 1 A simple
$n \times n$ multi-mesh network
with $n = 4$ (all links are not
shown)

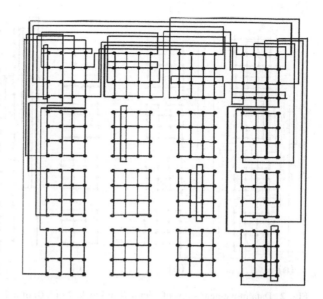

$\forall\ \beta,\ 0 \leq \beta \leq n - 1$, P $(\alpha, \beta, 0, y)$ are connected to P $(y, \beta, n - 1, \alpha)$, where $0 \leq y, \alpha \leq n - 1$, and inter block horizontal connection is identified by following rule.

$\forall\ \alpha,\ 0 \leq \alpha \leq n - 1$, P $(\alpha, \beta, x, 0)$ are connected to P $(\alpha, x, \beta, n - 1)$, where $0 \leq x, \beta \leq n - 1$.

4 Fourier Transformation Using Matrix–Vector Multiplication

Transformation of a matrix A of size $N \times N$ by a vector B of size N is given by $C = A \times B$, where C is the transformed vector of length N.

4.1 Parallel Implementation of Fourier Transform Using MM Network

The given matrix A is partitioned into $n \times n$ sub-matrices $A_{\alpha,\ \beta}$ where $0 \leq \alpha$, $\beta \leq n - 1$. In this way there are n^2 sub-matrices which are initially mapped to M (α, β), $0 \leq \alpha, \beta \leq n - 1$, each of which contains n^2 processors P $(\alpha, \beta, *, *)$. This '*' indicates all possible values from 0 to $n - 1$. Here each processor holds a single component of matrix A and all these values are generated in place according to the position of the node P (α, β, x, y), $0 \leq \alpha, \beta, x, y \leq n - 1$. For example,

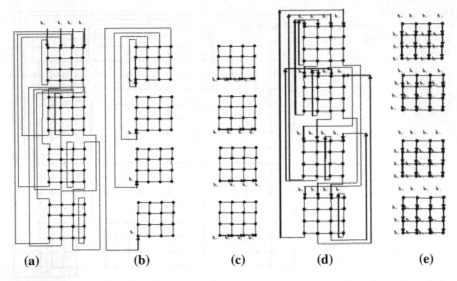

Fig. 2 Data movement steps of vector B in blocks M ($*$, 0) of a 4 × 4 MM: **a** Input through the *upper* boundary of the block M (0, 0). **b** Input values are moved to other blocks through *vertical inter block links*. **c** Propagation along row in each block. **d** Last row of data is moved along *vertical inter block links*. **e** Propagation along column direction within block

$\omega^{(\alpha n + x)(\beta n + y)}$ is the value of A at the node position P (α, β, x, y) and is calculated as $\cos\frac{(\alpha n+x)(\beta n+y)2\pi}{n} - i\sin\frac{(\alpha n+x)(\beta n+y)2\pi}{n}$ using the built in complex functionality $\omega = e^{\frac{2\pi i}{n}} = \cos\frac{2\pi}{n} - i\sin\frac{2\pi}{n}$. Next B values are input through the upper boundary of the MM network and are moved to all the other positions in the MM network. The movement of the B values are shown in Fig. 2 where only first four components of B vector are shown for the first column of blocks of a 4 × 4 Multi-Mesh.

Once the B values are populated and A values are calculated at the corresponding node, the next step towards completion of Fourier Transform is to perform multiplication at each node.

To calculate the $C_i \sum_{j=0}^{n^2-1} c_{ij}$ for each i, where $c_{ij} = a_{ij} \times b_j$, all c_{ij}'s are to be brought in a single block M (i/n, $i\%n$), as they are now scattered in ith rows of n different blocks. To achieve this n shift is performed along the horizontal inter block links of the MM network as shown in Fig. 3 given below.

Now the n^2 values of each mesh are summed along the column direction then horizontally along the first row of each block to make an element of the result vector C. After this step one more single step data movement is performed along the horizontal inter block links to bring the values of all the blocks M (α, 0), $0 \leq \alpha \leq n - 1$, at the left boundaries as shown in and Fig. 4. The final output is obtained at the left boundary of the MM network.

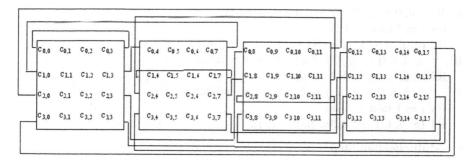

Fig. 3 Contents of blocks M (0, *) after n (=4) steps data movement along the *horizontal inter block links* in a 4×4 MM

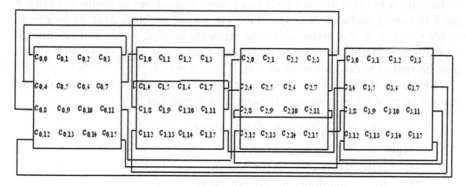

Fig. 4 Column sum followed by 0th row sum in each block of MM for $n = 4$ (*solid lines* with *arrowhead* indicates the direction of additions)

5 Parallel Discrete Fourier Transform Using MM Network

5.1 Algorithm PDFT

A. Initialization step

This step is subdivided into two parts. Register R1 s contain the initial values of matrix A, whereas registers R2 s contain the initial values of vector B.

Step 1: $\forall\ \alpha, \beta, x$ and y, $0 \le \alpha, \beta, x, y \le n - 1$ do in parallel
$\qquad$ R1 $(\alpha, \beta, x, y)\leftarrow$

Step 2: $\forall\ \beta$ and y, $0 \le \alpha, \beta, y \le n - 1$ do in parallel
$\qquad$ R2 $(0, \beta, 0, y)\leftarrow$

B. Propagation of B vector to the other rows of MM network

I. $\forall\ \beta$ and y, $0 \le \beta, y \le n - 1$ do in parallel
$\qquad$ R2 $(y, \beta, n, 0) \leftarrow$ R2 $(0, \beta, 0, y)$;

II. $\forall\ \alpha,\ \beta,\ 0 \leq \alpha,\ \beta \leq n - 1$ do in parallel
   for $i = 1$ to $n - 1$ do
   R2 $(\alpha,\ \beta,\ n - 1,\ i) \leftarrow$ R1 $(\alpha,\ \beta,\ n - 1,\ i - 1)$;
III. $\forall\ \alpha,\ \beta,\ i,\ 0 \leq \alpha,\ \beta,\ i \leq n - 1$ do in parallel
   R2 $(i,\ \beta,\ 0,\ \alpha) \leftarrow$ R2 $(\alpha,\ \beta,\ n - 1,\ i)$;
IV. $\forall\ \alpha,\ \beta,\ j,\ 0 \leq \alpha,\ \beta,\ j \leq n - 1$ do in parallel
   for $i = 1$ to $n - 1$ do
   R1 $(\alpha,\ \beta,\ i,\ j) \leftarrow$ R2$(\alpha,\ \beta,\ i - 1,\ j)$;

C. Multiplication Step

$\forall\ \alpha,\ \beta,\ x$ and $y,\ 0 \leq \alpha,\ \beta,\ x,\ y \leq n - 1$ do in parallel
R1 $(\alpha,\ \beta,\ x,\ y) \leftarrow$ R1 $(\alpha,\ \beta,\ x,\ y) \times$ R2 $(\alpha,\ \beta,\ x,\ y)$;

D. Data Movement Step

In the MM network horizontal inter block links form cycles of length $2n$ between the k th row of the block $M\ (i,\ j)$ and the j th row of the block $M\ (i,\ k)$ for $j \neq k$, $0 \leq i \leq n - 1$. For a given α, if data elements in $M(\alpha,\ *)$ are shifted through n positions along the horizontal cycles, then the ith row elements of $M(\alpha,\ j)$ will also be shifted to the jth row of $B(\alpha,\ i),\ 0 \leq \alpha \leq n - 1$.

/* Here, '*' indicates all possible values from 0 to $n - 1$, but the same value for it must be used on both sides of the assignment operator */

    $\forall\ \alpha$ and $\beta,\ 0 \leq \alpha,\ \beta \leq n\text{-}1$ do in parallel
    begin
    I.   R1$(\alpha,\ \beta,*,\ n\text{-}1) \leftarrow$ R1$(\alpha,*,\ \beta,\ 0)$;
    II.   for $j = n\text{-}1$ down to 1 do in parallel
        R1$(\alpha,\ \beta,*,\ j\text{-}1) \leftarrow$ R1$(\alpha,\ \beta,\ *,\ j)$;
        endfor
    end
    /* Steps I and II are in parallel */

E. Addition Step

    $\forall\ \alpha,\ \beta$ and $y,\ 0 \leq \alpha,\ \beta,\ y \leq n\text{-}1$ do in parallel
    begin
    for $i = n\text{-}1$ down to 1 do
    R1 $(\alpha,\ \beta,\ i\text{-}1,\ y) \leftarrow$ R1 $(\alpha,\ \beta,\ i\text{-}1,\ y) +$ R1 $(\alpha,\ \beta,\ i,\ y)$;
    /* R1 $(\alpha,\ \beta,\ 0,\ y)$ contain the column sums */
    for $j = 0$ to $n\text{-}2$ do
    R1 $(\alpha,\ \beta,\ 0,\ j + 1) \leftarrow$ R1 $(\alpha,\ \beta,\ 0,\ j+1) +$ R1 $(\alpha,\ \beta,\ 0,\ j)$;
    /* Summing along the 0th row in each block, the sum of the n^2 data values of the block is finally brought to the processor P $(\alpha,\ \beta,\ 0,\ n\text{-}1\ )$ */
    end

Table 1 Time complexities of PDFT computation in various architectures are given below

Architecture	Number of nodes	Length of vector	Time complexity
2D mesh [2]	N	N	$N + 2 N^{3/2}$
Multi dimentional mesh	N	N	$O\ (\log^2 N)$
Hypercube [6]	N	N	$O\ (logN)$
Binary tree [5]	$O\ (N)$	N	$O\ (NlogN)$
Multi mesh [7]	N	$N^{0.4}$	$O\ (N^{1/4})$
Star graph [6]	$n!$	$n!$	$O\ (n^2)$
Multi mesh (proposed)	N	$N^{1/2}$	$O\ (N^{1/4})$

F. Data Arrangement Step

In this step the output data vector C is moved to the 0th column of all the blocks $B\ (\alpha, 0), 0 \le \alpha \le n\text{-}1$.

$\forall\ \alpha$ and $\beta, 0 \le \alpha, \beta \le n\text{-}1$ do in parallel

R1 $(\alpha, 0, \beta, 0) \leftarrow$ R1 $(\alpha, \beta, 0, n\text{-}1)$;

/* It is a single step horizontal data movement along inter block links */

6 Time Complexity of Parallel DFT

The number of steps required for initialization stage A is 6 and for propagation stage is $2n$. Generation of partial products needs single step. Data movement stage D requires n steps. Addition stage E requires $2(n - 1)$. Data arrangement stage F is done in single step. So, total number of steps required for the entire process is $5n + 6$ or $5 N^{1/2} + 6$ where $N^2 = n^4$ which implies time complexity of this algorithm is $O\ (n)$ or $O\ (N^{1/2})$.

7 Comparative Study of Time Complexities of Parallel DFT in Multi-Mesh and Other Architectures

The time complexities of parallel DFT computation in various architectures have been tabulated in Table 1.

For the purpose of comparison, we are considering same mesh size $(n \times n)$ and same total number of nodes (n^4) for both MM and Multi-dimensional mesh. The diameters of MM and 4-D mesh are $2n$ and $4\ (n - 1)$ respectively, and degree of each node in MM is uniformly 4, whereas the node degree of each internal node is 8 for 4-D mesh. If smaller meshes are used to construct the multi-dimensional mesh, the node degree will increase. Moreover, the time complexity of proposed DFT computation in MM outperforms multi-dimensional mesh for practical sizes of meshes, i.e., mesh sizes less than 2048 $\times$ 2048.

In Hypercube better time complexity is achieved through its high node degree and number of links with increasing dimensions.

8 Conclusion and Future Work

A parallel algorithm for DFT of vector of length N $(=n^2)$ has been proposed with constant multiplication time and O $(\sqrt{N})$ addition and data communication time. Each processor of the MM network having N^2 $(=n^4)$ processors, generates a single component of the Fourier matrix depending on the processor position in MM network i.e. each of the components is generated in place.

The authors in [7] have implemented $n^{1.42} \times n^{1.42}$ matrix-by-matrix multiplication in O (n) time using n^4 processors (same as that used in the proposed work) of MM and indicated that DFT of vector length $n^{1.42}$ can be computed in the same way. In the present work, DFT of vector length n^2 is computed as linear transformation in O (n) time using n^4 processors of MM. The Fourier matrix is calculated in-place at each node of the MM, so there is no need to input the Fourier matrix through the left boundary of each mesh (block) as was done in [7], and vector to be transformed was input only through the upper boundary of the MM whereas the input was made through the upper boundary of each mesh (block) in [7].

As a future work, we may consider a generalized MM network where total number of processors in MM network is $m^2 \times n^2$ having $m \times n$ number of processors in each block of the MM network for $m, n \geq 3$.

References

1. Cooley, J.M., Tukey, J.W.: An algorithm for the machine computation of the complex Fourier series. Math. Comput. **19**, 297–301 (1965)
2. Strong, J.P.: The Fourier transform on mesh connected processing arrays such as massively parallel processors. CAPAIDM **19**, 190–196 (1986)
3. Shousheng, H., Torkelson, M.: A systolic array implementation of common factor algorithm to compute DFT. In: International Symposium on Parallel Architectures, Algorithms and Networks (ISPAN), pp. 374–381 (1994)
4. Sinha, B.P., Mukherjee, A.: Parallel sorting algorithm using multiway merge and its implementation on a multi-mesh network. J. Parallel Distrib. Comput. **60**, 891–960 (2000)
5. Jaja, J.: An Introduction to Parallel Algorithms. Addison-Wesley (1992)
6. Fragopoulou, P., Akl, S.G.: A parallel algorithm for computing Fourier transform on the star graph. Trans. Parallel Distrib. Syst. **5**(5), 525–531 (1994)
7. Das, D., De, M., Sinha, B.P.: A new network topology with multiple meshes. IEEE Trans. Comput. **48**(5), 536–551 (1999)
8. De, M., Das, D., Sinha, B.P.: An efficient sorting algorithm on the multi-mesh network. IEEE Trans. Comput. **46**(10), 1132–1137 (1997)
9. Sinha, B. P.: Multi-mesh an efficient topology for parallel processing. In: Proceedings of 9th International Parallel Processing Symposium, (IPPS), pp. 17–21 (1995)

Dynamic Checkpoint Data Replication Strategy in Computational Grid

Ramesh Babu and Subba Rao

Abstract Computational grid is a good solution to large scale data processing and management problems including efficient checkpoint transfer and replication. Due to heterogeneous nature of grids most of time grids more prone to failure or latency delay. Subsequently checkpointing and replication is indispensable to tolerate such faults efficiently. Dynamic checkpoint data replication in computational grid aims to improve data access time and to utilize network and storage resources efficiently. Since the data checkpoints are very large and grid storages are limited, managing replicas in storage for the purpose of more effective utilization of them require more attention. In this work, a dynamic checkpoint data replication mechanism is proposed, which is called checkpoint based optimal replication (CBOR). CBOR selects a checkpoint for replication and calculates a suitable number of copies and grid sites for replication by setting different weight for each data access record. The data access records in the near past have higher weights. A grid simulator Optorsim is used to evaluate the performance of CBOR dynamic replication strategy. The experimental results show that CBOR successfully increases the effective network usage by finding out a popular checkpoint and replicates it to a suitable site.

Keywords Computational grid · Dynamic checkpoint · Optimal replication · Job scheduling · Node failure

R. Babu (✉)
Jawaharlal Nehru Technological University, Kakinada, AP, India
e-mail: crb.challagundla@gmail.com

S. Rao
Sri Venkateswara University, Tirupati, AP, India

G. P. Biswas and S. Mukhopadhyay (eds.), *Recent Advances in Information Technology*,
Advances in Intelligent Systems and Computing 266, DOI: 10.1007/978-81-322-1856-2_11,
© Springer India 2014

1 Introduction

Grid technology is modern computing approach for solving very high computational tasks with optimal utilization of resources in large scale heterogeneous systems. Mathematical models of problems from science and engineering and many applications require maximum usage heterogeneous resources together. As with any such field, definitions and standards have yet to settle into fixed forms, with many very diverse projects all claiming to be grids. It is therefore necessary to consider the fundamental concepts of a grid and come to some definition of what is, and what is not, a grid. As a starting point, this section will consider the simple definition of a grid as a computing utility—the aspiration of many computer scientists. Then, the technological prerequisites for the execution of such an aspiration are described. Therefore a well-known definition of a grid is adopting to data grids in particular for this replication. Most of the time grids are more prone to failures or delay in executing tasks, because diversity of applications requires different computational resources. So in order to tackle this theme fault tolerance come to existence. Commonly adopting techniques for fault tolerance is periodic job checkpointing and replication. Checkpointing is a system state over a time of execution for subsequent usage and restoration need. Typically checkpoint data are stored in stable storage such as data grids [9].

The checkpointing and replication strategies aim to find out the popular checkpoint from the information about the number of accesses for checkpoints and discuss how to place a new replica. It may be reasonable to determine the popular checkpoint according to the number of accesses. A higher number of accesses file gets a higher popularity. However this kind of methods in terms of access times will have a problem about valid period of historical records. If a checkpoint file was accessed for a lot of times in the past, while there was almost none recently, the checkpoint would still be created because the number of accesses for the checkpoint is larger than other checkpoints [3].

Dynamic replication is an approach of placement of replicas based on popularity of the data. The method had been designed around a hierarchical model. When the number of hits for a specific dataset exceeds the replication threshold it triggers the creation of a replica on the server that directly serves the user's client. To take advantage of the hierarchical topology the client looks up checkpoints from the client to the root and the root creates replicas of the request checkpoint at every node which is on the path from the client to the root. Hence the access time requires replicating checkpoint data could radically reduced. So disadvantage of this technique is a bottleneck of storage space. In conclusion, the storage space and access latency will be the trade off in designing replication strategy [5].

In this work dynamic replication strategy called checkpoint based optimal replication (CBOR) is proposed which solves the problem of valid period of historical records by giving different weights to records having different ages. All records whether new or old, contribute to determine a popular checkpoint that should be replicated.

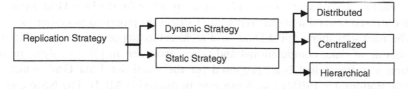

Fig. 1 Classification of replication

2 Computational Replication

2.1 Classification of Data Replication Strategies

The motivation for replication is how to enhance data availability, accessibility, reliability, and scalability. Generally, replication algorithms are either static or dynamic as shown in Fig. 1. In static approaches the created replica will exist in the same place till user deletes it manually or its duration is expired. The disadvantages of the static replication strategies, which may not adapt to changes in user behaviour and they are not appropriate for huge amount of data and large number of users. Certainly static replication methods have some advantages such as they do not have the overhead of dynamic algorithms and job scheduling. On the other hand, dynamic strategies create and delete replicas according to the changes in Grid environments, i.e. users' checkpoint access pattern. As Data Grids are dynamic environments and the requirements of users are variable during the time, dynamic replication is more appropriate for these systems. But many transfers of huge amount of data that are a consequence of dynamic algorithm can lead to a strain on the network's resources. So, inessential replication should be avoided. A dynamic replication scheme may be implemented either in a centralized or in a distributed approach. These methods also have some drawbacks such as overload of central decision centre further grows if the nodes in a data grid enter and leave frequently. In case of the decentralized manner, further synchronization is involved making the task hard.

2.2 Related work

The drawback of transferring a checkpoint from one site to another in real time is bandwidth consumption and access delay. Data replication technique is a frequently reduce bandwidth consumption and access latency by taking one or more checkpoints to other sites. Two kinds of replication methods are possible. The static replication creates and manages the replicas manually. The dynamic replication changes the location of replicas and creates new replicas to other sites automatically. Judging from the fact that resources or data checkpoints are always changing, it is clear that dynamic replication is more appropriate for the Data Grid. In this we have observed several dynamic replication strategies proposed in earlier [9].

The multi-tier hierarchical Data Grid architecture supports an efficient method for sharing of data, computational and other resources. In [3], two dynamic replication algorithms had been proposed for the multi-tier Data Grid, which are Simple Bottom–Up (SBU) and Aggregate Bottom-Up (ABU). The basic concept of SBU is to create the replicas as close as possible to the clients that request the data checkpoints with a rate exceeding a predefined threshold. If the number of requests for checkpoint f exceeds the threshold and f has existed in the parent node of the client which has the highest request rate, then there is might not to replicate. The SBU replication algorithm has the disadvantage that it does not consider the relations among these historical records but processes the records individually. After aggregation, the checkpoint with the highest request rate will be replicated if the rate is above a threshold. With the exception of aggregation steps, all concepts in ABU are similar to the SBU replication algorithm.

The Fig. 2 shows historical records checkpoint file data C1, C2, C3… and associated process_Ids in different grid sites X, Y, and Z etc. Here p1, p3, p6 have used the replica file due to failure of process or unavailability of resources. A centralized dynamic replication mechanism was proposed in [4, 2]. It determines the popular checkpoint the same as ABU does by analyzing the data access history. In addition, a special idea was proposed to find out the average of all number of accesses (NOA) for records in the history table, which acts as entry to select the popular data checkpoints.

$$\overline{NOA} = \sum_{h \in H} NOA(h)$$

where |H| indicates the number of data checkpoints that has been requested. NOA (h) is the content of hth record in the history table H, and $\overline{NOA}$ field represents the number of accesses for a checkpoint. Only the checkpoints with NOA exceeding NOA will be replicated. After finding the NOA, the historical records whose NOA values are less than NOA will be separated [7].

3 Checkpoint Replication Design

In this work, we propose a dynamic checkpoint based optimal replication strategy. We consider the grid sites as a heterogeneous cluster. There is a cluster scheduler used to supervise the site information in a cluster. A cluster scheduler connects the checkpoint access information with another scheduler. It then determines the checkpoint that should be replicated and the place to be replicated. Each site maintains a detailed access chronological checkpoint record for each checkpoint.

Fig. 2 An example of the history and the node relation

C3	p2, p4, p7,...	7	Z
...	...	...	..

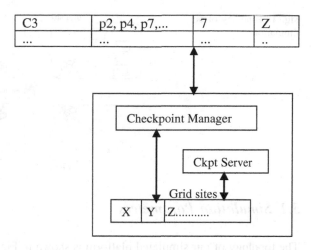

Fig. 3 Checkpoint strategy

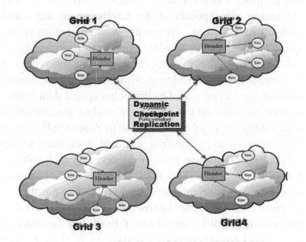

Checkpoint Replication Strategy

The record is stored in the form of time_Interval, checkpoint_Id, cluster_Id, which indicates the checkpoint field has been accessed by a site located in the cluster cluster_Id at timestamp. Often each site sends its records to the cluster scheduler records the same cluster which will be aggregated and summarized by the cluster scheduler. Figure 3 is an example to show the records of a cluster scheduler and their aggregation. The aggregated record should be the same in each cluster if the information is exchanged completely.

Fig. 4 Topology of our simulation

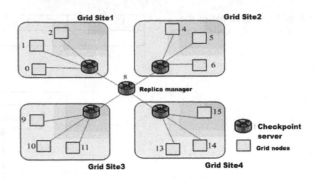

3.1 Simulation Parameters

The topology of our simulated platform is shown in Fig. 4. There are four clusters and each one has three sites. Nodes 16 have the most capacity in order to hold all the master checkpoints at the beginning of the simulation. The others have a uniform size, 80 GB. All the network bandwidth is set as 500 mbps. The storage space at each site is 80 GB. The connection bandwidth is 500 mbps. There are 1000 jobs with 10 different access patterns. Each data checkpoint to be accessed is 500 mega bytes. In order to simplify the requirements, we do not consider the consistency of replicas. Optimal checkpoint data replication strategies commonly assume that the data is read-only in data grid environments. We have simulated with 1000 jobs. A job is submitted to Resource Broker every 60 s. Resource broker then submits to computing element according to (Queue Access Cost) QAC scheduling. There are four job types, and each job type requires specific checkpoints for execution. The order of checkpoints accessed in a job is sequential and is set in the job configuration checkpoint. The number of checkpoints in our simulation is 150, and a checkpoint size is 500 MBs. To demonstrate the advantages of the dynamic replica algorithm, CBOR will be compared with Simple Optimizer and LFU (Least Frequently Used), LALW (Latest Accesses Largest Weight). The Simple Optimizer is a base case, which no replication and checkpoints are accessed remotely. The LFU, LALW algorithm always replicates, deleting those checkpoints least frequently accessed if the space of SE is not enough for replication [4, 10] and [11].

4 Simulation Framework and Results

4.1 Simulation Inputs

As well as the basic architecture, the detailed conceptual model of the grid applications and fabric components is crucial in achieving a valid simulation.

These are components which must be easily configurable for each simulation and so are classified together here as simulation inputs.

4.1.1 Grid Topology

A grid site may have a Computing Element (CE), a Storage Element (SE) or both. Each site also has a Replica Optimiser (RO) which makes decisions on replications to that site. The topology of the grid—the number of sites and the way they are connected—has a strong impact on grid behaviour. It is therefore important that Optorsim take any grid topology as input, i.e. that the user can specify the storage capacity and computing power at each site, and the capacity and layout of the network links between each. SEs is defined to have a certain capacity, in MB, and CEs to have a certain number of "worker nodes" with a given processing power. Sites which have neither a CE nor an SE act as routers on the network.

4.1.2 Jobs and Checkpoints

A job usually processes a certain number of checkpoints. This is simulated in Optorsim by defining a list of jobs and the checkpoints that they need, with their sizes. The list of checkpoints that a job may need is defined to be its dataset. A job, when sent to a CE, will process some or all of the checkpoints in its dataset, according to the access pattern which has been chosen. The time a checkpoint takes to process depends on its size and on the number and processing power of worker nodes at the CE [7, 8].

4.2 Simulation Parameters

There are number of parameters other than the simulation inputs, which are of interest when performing simulations of a data grid. The most important are outlined here a full description of all the parameters is given in the following [6]. These are:

- **Initial Checkpoint Distribution:** The distribution of checkpoints around the grid at the start of the simulation can affect the behaviour of the replica optimisation strategies. A set of the master checkpoints is therefore placed at designated sites, and the option is also given to fill up the sites with replicas of these checkpoints before the simulation gets under way.
- **Access Patterns:** Different kinds of job may access the checkpoints in the dataset in a different way. To simulate this, the checkpoints in the dataset are ordered. Some jobs may process each checkpoint in sequence; others may miss some checkpoints or read them in a different order.

Table 1 Packages in optorsim associated functionality

Package	Scope
Optorsim	Overall simulation control
	Resource broker
	Users
	Simulation out graphical user interface
Optorsim. optor	Replica optimization
Optorsim. Time	Control of timing
Optorsim. Infrastructure	Grid fabric (sites, network etc.)
Optorsim reptorsim	Replica management and cataloguing
Optorsim. auctions	Auction functionality fro economic models

- **Users:** The pattern and rate of job submission by users may affect the behaviour of job queues at sites and hence the behaviour of the Resource Broker and Replica Optimisers.
- **Number of Jobs:** Optimisation strategies may perform differently under conditions when the grid is under-utilized (few jobs) or congested (many jobs) so it is important to vary the number of jobs and measure this effect.
- **Job Scheduler:** The algorithm used by the Resource Broker for its scheduling decisions will clearly have an important effect on the running of the jobs and performance of the grid.
- **Optimiser:** The algorithm used by the Replica Optimisers for the checkpoint replication strategy, which is the original research focus of Optorsim.

4.3 Implementation

The code is structured into several packages, each of which deals with a different part of the simulation. The list of packages and their main remit is shown in Table 1. Details of most of the classes within these packages are not given here, but can be found in the Java doc API which comes with the Optorsim 2.1 release.

4.4 Run-Time Processes

The simulation inputs are read in at the start of a simulation run from a set of configuration checkpoints, one of which describes the grid topology, one which describes the jobs, checkpoints and site policies, one which contains information on the network bandwidth variation and one which contains all the simulation parameters. When the sites and grid have been instantiated with the given inputs,

Fig. 5 Effective network usage of grid

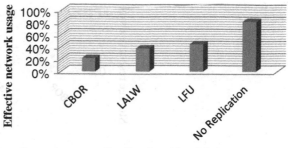

Replication Algorithms

the Computing Element and Resource Broker threads are started, along with the simulation timing thread and followed by the Users thread [1, 2].

5 Experimental Results

An evaluation metric effective network usage *(enu)* in Optorsim 2.1 is used to estimate the usage of grid resource. It is defined as follows.

$$enu = \left(n_{remotecheckpoint} + n_{replication}\right)/n_{checkpoint}$$

where, n remote checkpoint accessed is the number of accesses that CE reads a checkpoint from a remote site, n- checkpoint replication is the total number of checkpoint replication occurs, and N checkpoint accesses the number of times that CE reads a checkpoint from a remote site or reads a checkpoint locally. A lower value indicates that the utilization of network bandwidth is more efficient.

Here Fig. 5 depicts the comparison of three replication strategy for ENU. Simple Optimizer has 100 % ENU because CEs read all checkpoints from remote sites. The CBOR and LFU, LALW algorithms can improve ENU about 40–50 %. Moreover, the ENU of CBOR is lower by 16 % compared to LFU, LALW algorithms. The reason is that LFU, LALW algorithms always replicates, so that the large value of N-checkpoint replication will increase the ENU. In contrast CBOR pre-replicates checkpoints that may be accessed by next job to the SE of site regularly. The CBOR surpassed LFU, LALW algorithms in ENU.

Here Fig. 6 illustrates storage resource usage, which is the percentage of available spaces that are used. It depends on the number of replica, thus, the storage resource usage of LFU, LALW algorithm is higher than Simple Optimizer and CBOR. Simple Optimizer has the smallest value because it always read requested checkpoint remotely.

The mean job execution time for various strategies is shown in Fig. 7. However, LFU, LALW and CBOR show similar performance.

Fig. 6 Storage resources usage

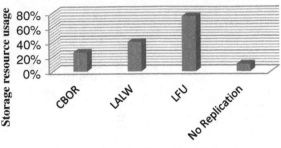

Replication Algorithms

Fig. 7 Mean job execution time

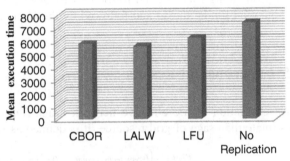

Replication Algorithms

6 Conclusions

In this paper, we propose a dynamic replication strategy called Checkpoint based optimal replication. At intervals, the dynamic replication algorithm collects the data access history, which contains checkpoint name, the number of requests for checkpoint, and the sources that each request came from. In order to evaluate the performance of our dynamic replication strategy, we use the Grid simulator Optorsim to simulate a realistic Grid environment. The simulation results show that the average job execution time of CBOR is similar to LFU, LALW optimizer, but excels in terms of Effective Network Usage.

For the future work, we will try to further reduce the job execution time. There are two factors to be considered. One is the length of a time interval. If the length is too short, the information about data access history is not enough. Then information about data access history will be more important in contributing to find the popular optimal checkpoints.

References

1. Bell, W.H., Cameron, D.G., Capozza, L., Millar, P., Stockinger, K., Zini, F.: Optorsim—a grid simulator for studying dynamic data replication strategies. Int. J. High Perform. Comput. Appl. **17**(4), 403–416 (2003)
2. Cameron, D.G., Schiaffino, R.C., Millar, P., Nicholson, C., Stockinger, K., Zini, F.: OptorSim: a grid simulator for replica optimisation. In: UK e-Science All Hands Conference, 31 August—3 Sept 2004
3. Chang, R.-S., National Dong Hwa University, Hualien, Chang, H.-P., Wang, Y.-T.: A dynamic weighted data replication strategy in data grids, In: Computer Systems and Applications, AICCSA 2008, IEEE/ACS 2008
4. Chervenak, A., Foster, I., Kesselman, C., Salisbury, C., Tuecke, S.: The data grid: towards an architecture for the distributed management and analysis of large scientific datasets. J. Netw. Comput. Appl. **23**, 187–200 (2000)
5. Dilli babu, S., Ramesh Babu, C., Subba rao, C.D.V.: An efficient fault-tolerance technique using check-pointing and replication in grids using data logs. In: publications of problems and application in engineering research—paper, vol 04. special issue 01, 2013
6. Foster, I.: Globus toolkit version 4: software for service-oriented systems, In: IFIP International Conference on Network and Parallel Computing, vol. LNCS 3779, pp. 2–13. Springer-Verlag (2005)
7. Hoschek, W., Jaen-Martinez, F.J., Samar, A., Stockinger, H., Stockinger, K.: Data management in an International data grid project. In: Proceedings of the First IEEE/ACM International Workshop on Grid Computing (GRID '00), Bangalore, India, Dec 2000. Lecture Notes in Computer Science, vol. 1971, pp 77–90
8. Ranganathan, K., Foster, I.: Identifying dynamic replication strategies for a high-performance data grids, In: Proceeding of 3rd IEEE/ACM International Workshop on Grid Computing, Denver, Nov 2002. Lecture Notes on Computer Science, vol. 2242, pp. 75–86. Springer, Berlin (2002)
9. Singh, A.K., Srivastava S., Shanker, U.: A survey on dynamic replication strategies for improving response time in data grids, IJBSTR, July 2013
10. Tang, M., Lee, B.-S., Tang, X., Yeo, C.-K.: The impact of data replication of job scheduling performance in the data grid. Future Gener. Comput. Syst. **22**, 254–268 (2006)
11. The European Data Grid Project. http://eudatagrid.web.cern.ch/eu-datagrid/Winter

Part IV
Technical Session-IV: Mathematical Computing

Estimation of Population Mean in Presence of Non Response in Two-Occasion Successive Sampling

Arnab Bandyopadhyay and Garib Nath Singh

Abstract The present work intended to emphasize the role of one varying auxiliary variable at both the occasions to improve the precision of estimates of population mean at current occasion in two-occasion successive sampling in presence of non-response. Two different efficient estimators are proposed and their theoretical properties are examined. The proposed estimators have been compared with (i) the sample mean estimator in the presence of non response, where no past information is used and (ii) the estimator suggested by Singh and Priyanka [1], which is a linear combination of the means of the matched and unmatched portions of the sample at the current occasion [1]. Theoretical results have been interpreted through empirical studies which are followed by suitable recommendations.

Keywords Non-response · Successive sampling · Regression · Exponential · Chain type estimator · Bias · Mean square error · Optimum replacement strategy

1 Introduction

In many situations of social, demographic, industrial and agricultural surveys, the same population is surveyed repeatedly and the same variable is measured on each occasion, so that development over time may be followed. In such studies, sampling over successive occasions (repetitive sampling) plays an important role to provide the reliable and the cost effective estimates of the real life situations at

A. Bandyopadhyay (✉)
Department of Mathematics, Asansol Engineering College, Asansol 713305, India
e-mail: arnabbandyopadhyay4@gmail.com

G. N. Singh
Department of Applied Mathematics, Indian School of Mines, Dhanbad 826004, India
e-mail: gnsingh_ism@yahoo.com

G. P. Biswas and S. Mukhopadhyay (eds.), *Recent Advances in Information Technology*, 109
Advances in Intelligent Systems and Computing 266, DOI: 10.1007/978-81-322-1856-2_12,
© Springer India 2014

different successive points of time. Theory of successive sampling appears to have started with the work of Jessen [2]. Further the theory of successive (rotation) sampling was extended by many authors including Singh and Priyanka [1], Singh and Karna [3], Singh and Homa [4]. In successive sampling, it is common practice to use the information collected on a previous occasion as auxiliary variable to improve the precision of the estimates at current occasion different period of time. However, it is common experience in sample surveys that data cannot always be collected from all the units selected in the sample. For example, the selected families may not be at home at the first attempt and some may refuse to cooperate with the interviewer even if contacted. As many respondents do not reply, available sample of returns is incomplete. The resulting incompleteness is called non-response and is sometimes so large that can completely vitiate the results. Hansen and Hurwitz suggested a sub-sampling technique of handling non-response in such situations [5]. Singh and Priyanka [1] and Singh and Karna [3] used the Hansen and Hurwitz [5] technique for estimation of population mean at current occasion in the context of successive sampling over two occasions. In two occasions successive sampling, a portion of sample is matched from the previous occasion and it is assumed that whole units respond at first occasion. So, we may think that as they are familiar with the questionnaire at first occasion, therefore, they may not have any hesitation in responding at the second occasion for the units in the matched portion of the sample. At the current occasion a sample is drawn afresh from the remaining units, so there may be possibility of non-response at current occasion. Motivated with the above arguments and following Hansen and Hurwitz [5] technique, we have proposed two estimators to estimate the population mean in presence of non-response at current occasion in two occasions successive sampling. Relative comparisons of efficiencies of the proposed estimators with some contemporary estimators of population mean defined for the same situation are discussed. Empirical studies show highly significant gains for the proposed estimators which are followed by suitable recommendations.

2 Proposed Estimators

Consider a finite population $U = (U_1, U_2, U_3, \ldots, U_N)$ of N units which has been sampled over two occasions. The character under study be denoted by x (y) on the first (second) occasions respectively. It is assumed that information on an auxiliary variable z_1 (z_2) whose population mean is known, is available on the first (second) occasion respectively. Let a simple random sample (without replacement) of size n be selected on the first occasion and a random sub-sample from the sample selected on the first occasion of $m = n\lambda$ units is retained (matched) for its use on the second occasion. We assume that non-response occurs only at the current occasion, so that the population can be divided into two classes, those who will respond at the first attempt and those who will not. Let the sizes of these two classes be N_1 and N_2 respectively. Now, at the second (current) occasion a simple random sample (without replacement) of $u = (n-m) = n\mu$ units is drawn afresh from the entire population so that the sample size on the second (current) occasion is also n. λ and μ $(\lambda + \mu = 1)$ are the fractions of matched and fresh samples respectively at the second (current) occasion. We assume that in the fresh sample on the second (current) occasion u_1 units respond and u_2 units do not respond. Let $u_{2h}(u_{2h} = u_2/k, k > 1)$ denote the size of sub sample drawn from the non-response class in the fresh sample on the current (second) occasion. The following notations are considered for the further use:

$\overline{X}, \overline{Y}, \overline{Z}_1, \overline{Z}_2$	Population means of the variables x, y, z_1 and z_2 respectively.
$\overline{x}(n), \overline{x}(m), \overline{y}(u_1), \overline{y}(u_{2h}),$ $\overline{y}(m), \overline{z}_1(n), \overline{z}_1(m), \overline{z}_2(m),$ $\overline{z}_2(u_1), \overline{z}_2(u_{2h})$	Sample means of the respective variables based on the sample sizes shown in braces.
$\rho_{yx}, \rho_{yz_1}, \rho_{yz_2}, \rho_{xz_1}, \rho_{xz_2}, \rho_{z_1z_2}$	Correlation coefficients between the variables shown in suffice
$S_x^2, S_y^2, S_{z_1}^2, S_{z_2}^2$	Population mean squares of the variables x, y, z_1 and z_2 respectively
$C_y, C_x, C_{z_1}, C_{z_2}$	Coefficients of variation of the variables shown in suffice
$\beta_{yx}, \beta_{xz_1}, \beta_{yz_2}$	Population regression coefficients between the variables shown in suffice
$W_2 = \dfrac{N_2}{N}$	The proportion of non-response units in the population

To estimate the population mean $\overline{Y}$ at the second (current) occasion, two different sets of estimators are considered. One set of estimators $S_u = \{T_u\}$ based on sample of size u $(= n\mu)$ drawn afresh on the second occasion and the second set of estimators $S_m = \{T_{1m}, T_{2m}\}$ based on the sample of size m $(=n\lambda)$ common with both the occasions. Motivated by the well known regression estimators and influencing with the Hansen and Hurwitz [5] technique, we suggest the estimators T_u as

$$T_u = \overline{y}_u^* + b_{yz_2}^* \left(\overline{Z}_2 - \overline{z}_2^*(u)\right) \tag{1}$$

where $b_{yz_2}^* = \dfrac{S_{yz_2}^*}{S_{z_2}^{*2}}$ is the estimate of the population regression coefficient β_{yz_2} of y on $\overline{z}_2$, $\overline{y}^*$ and $\overline{z}_2^*$ are the Hansen–Hurwitz estimators of the population means $\overline{Y}$ and $\overline{Z}_2$ respectively and are defined by $\overline{y}^* = \dfrac{u_1\overline{y}(u_1) + u_{2h}\overline{y}(u_{2h})}{u}$, $\overline{z}_2^* = \dfrac{u_1\overline{Z}_2(u_1) + u_{2h}\overline{Z}_2(u_{2h})}{n}$, $S_{yz_2}^*$ and $S_{z_2}^{*2}$ are the population covariance between the variables (y, z_2) and the population mean square of z_2 respectively in the non-response part of the population.

Similarly, motivated by the exponential estimator suggested by Bahl and Tuteja [6], we define the following estimators of set S_m as

$$T_{1m} = \overline{y}_m^1 + b_{yx}(m)(\overline{x}_n^1 - \overline{x}_m^1) \tag{2}$$

and

$$T_{2m} = \overline{y}_m^2 + b_{yx}(m)(\overline{x}_n^2 - \overline{x}_m^2) \tag{3}$$

where

$$\overline{y}_m^1 = \overline{y}(m) + b_{yz_2}(m)(\overline{Z}_2 - \overline{z}_2(m)), \ \overline{x}_n^1 = \overline{x}(n) + b_{xz_1}(n)(\overline{Z}_1 - \overline{z}_1(n))$$

$$\overline{x}_m^1 = \overline{x}(m) + b_{xz_1}(m)(\overline{Z}_1 - \overline{z}_1(m)), \ \overline{x}_n^2 = \overline{x}(n)\exp\left(\frac{\overline{Z}_1 - \overline{z}_1(n)}{\overline{Z}_1 + \overline{z}_1(n)}\right),$$

$$\overline{x}_n^2 = \overline{x}(n)\exp\left(\frac{\overline{Z}_1 - \overline{z}_1(n)}{\overline{Z}_1 + \overline{z}_1(n)}\right), \ \overline{x}_m^2 = \overline{x}(m)\exp\left(\frac{\overline{Z}_1 - \overline{z}_1(m)}{\overline{Z}_1 + \overline{z}_1(m)}\right)$$

and $b_{yx}(m), b_{yz_2}(m), b_{xz_1}(n), b_{xz_1}(m)$ are the sample regression coefficients between the variables shown in suffice and based on the sample sizes indicated in the braces.

Combining the estimators of sets S_u and S_m, we have the final estimators of population mean $\overline{Y}$, at second (current) occasion as

$$T_i = \varphi_i T_u + (1 - \varphi_i) \ T_{im}(i = 1, 2) \tag{4}$$

where $\varphi_i \ (i = 1, 2)$ are the real constants $(0 \leq \varphi_i \leq 1)$.

3 Optimum Replacement Strategies for the Proposed Estimators $T_i(i = 1, 2)$

Since, T_u and $T_{im} \ (i = 1, 2)$ are chain-type regression and regression–exponential estimators, they are biased estimators of population mean and the combined estimators $T_i(i = 1, 2)$ defined in Eq. (4) depend on the different values of $\varphi_i(i = 1, 2)$. Therefore, proceeding as Singh and Priyanka [1] and Singh and Karna [3] the minimum mean square errors Min. M(.) of our proposed estimators $T_i(i = 1, 2)$ to the first order of approximation are obtained as

Min. $M(T_1)$

$$= \frac{\mu_1^2(b_1 - fc_1)f \ B + \mu_1\{(b_1 - fc_1)A + f \ B(a_1 - b_1 + fc_1)\} + A(a_1 - b_1 + fc_1)}{\mu_1^2(b_1 - fc_1 - f \ B) + \mu_1(a_1 - b_1 + fc_1 - A + f \ B) + A} \frac{S_y^2}{n} \tag{5}$$

and

Min. $M(T_2)$

$$= \frac{\mu_2^2(b_2 - fc_2)f\ B + \mu_2\{(b_2 - fc_2)A + f\ B(a_2 - b_2 + fc_2)\} + A(a_2 - b_2 + fc_2)}{\mu_2^2(b_2 - fc_2 - f\ B) + \mu_2(a_2 - b_2 + fc_2 - A + f\ B) + A} \frac{S_y^2}{n}$$

(6)

respectively,

where

$$A = \{1 + W_2(k-1)\}\left(1 - \rho_{yz_2}^2\right), \quad B = -\left(1 - \rho_{yz_2}^2\right),$$

$$a_1 = \left(1 - \rho_{yz_2}^2\right) + \rho_{yx}\left\{\rho_{yx}\left(1 - \rho_{yz_1}^2\right) + 2\left(\rho_{yz_1}^2 - \rho_{yx} - \rho_{yz_2}\rho_{yz_1}\rho_{z_1z_2} + \rho_{yz_2}^2\right)\right\}, \quad f = \frac{n}{N},$$

$$b_1 = \rho_{yx}\left\{\rho_{yx}\left(1 - \rho_{yz_1}^2\right) + 2\left(\rho_{yz_1}^2 - \rho_{yx} - \rho_{yz_2}\rho_{yz_1}\rho_{z_1z_2} + \rho_{yz_2}^2\right)\right\}, \quad c_1 = -\left(1 - \rho_{yz_2}^2\right),$$

$$a_2 = \left(\frac{5}{4} - \rho_{yz_2}\right) + \rho_{yx}\left[\rho_{yx}\left(\frac{5}{4} - \rho_{yz_1}\right) + \rho_{yz_2} + \rho_{yz_1} - 2\left(\rho_{yx} + \frac{1}{4}\rho_{z_1z_2}\right)\right],$$

$$\text{and } b_2 = \rho_{yx}\left[\rho_{yx}\left(\frac{5}{4} - \rho_{yz_1}\right) + \rho_{yz_2} + \rho_{yz_1} - 2\left(\rho_{yx} + \frac{1}{4}\rho_{z_1z_2}\right)\right], \quad c_2 = -\left(\frac{5}{4} - \rho_{yz_2}\right).$$

and μ_i $(i = 1, 2)$ are the fraction of samples to be drawn afresh on the second (current) occasion.

Now, to determine the optimum values μ_i $(i = 1, 2)\{$say $\mu_i^{(o)}$ $(i = 1, 2)\}$ so that population mean $\overline{Y}$ may be estimated with maximum precision we further minimize Min. $M(T_i)$ $(i = 1, 2)$ in the Eqs. (5) and (6) and we have the following optimal values of them as

$M(T_1^o)_{opt}$

$$= \frac{\mu_1^{(o)2}(b_1 - fc_1)f\ B + \mu_1^{(o)}\{(b_1 - fc_1)A + fB(a_1 - b_1 + fc_1)\} + A(a_1 - b_1 + fc_1)}{\mu_1^{(o)2}(b_1 - fc_1 - f\ B) + \mu_1^{(o)}(a_1 - b_1 + fc_1 - A + f\ B) + A} \frac{S_y^2}{n}$$

(7)

and

$M(T_2^o)_{opt}$

$$= \frac{\mu_2^{(o)2}(b_2 - fc_2)f\ B + \mu_2^{(o)}\{(b_2 - fc_2)A + f\ B(a_2 - b_2 + fc_2)\} + A(a_2 - b_2 + fc_2)}{\mu_2^{(o)2}(b_2 - fc_2 - f\ B) + \mu_2^{(o)}(a_2 - b_2 + fc_2 - A + f\ B) + A} \frac{S_y^2}{n}$$

(8)

Respectively.

Table 1 Optimum values of μ_1 and percent relative efficiencies of T_1

ρ_{yz_1}		0.5			0.7			0.9		
ρ_{yz_2}	$\rho_{z_1z_2}$	$\mu_1^{(o)}$	$E_1^{(1)}$	$E_1^{(2)}$	$\mu_1^{(o)}$	$E_1^{(1)}$	$E_1^{(2)}$	$\mu_1^{(o)}$	$E_1^{(1)}$	$E_1^{(2)}$
ρ_{yx}		*0.5*								
0.5	0.5	0.62	121.63	122.43	0.57	116.71	117.48	0.53	110.59	111.32
	0.7	0.64	123.70	124.53	0.59	119.27	120.06	0.55	113.35	114.10
	0.9	0.67	125.88	126.72	0.62	122.03	122.84	0.57	116.35	117.13
0.7	0.5	0.53	163.12	164.20	0.50	156.56	157.60	0.47	148.03	149.01
	0.7	0.55	167.83	168.94	0.53	162.36	163.43	0.49	154.19	155.21
	0.9	0.58	173.053	174.20	0.56	168.98	170.10	0.52	161.36	162.43
0.9	0.5	0.38	323.51	325.65	0.36	309.99	312.04	0.34	291.13	293.06
	0.7	0.40	339.26	341.51	0.38	329.50	331.69	0.36	311.68	313.75
	0.9	0.42	357.99	360.36	0.41	353.96	356.30	0.40	338.13	340.37
ρ_{yx}		*0.7*								
0.5	0.5	0.70	128.14	110.03	0.62	121.69	104.50	0.55	113.68	*
	0.7	0.79	131.28	112.73	0.67	125.86	108.08	0.58	118.02	101.35
	0.9	1	133.33	114.49	0.76	130.30	111.89	0.63	122.97	105.59
0.7	0.5	0.55	168.12	144.36	0.52	160.24	137.60	0.48	149.90	128.73
	0.7	0.59	175.63	150.82	0.56	169.33	145.40	0.51	159.24	136.74
	0.9	0.66	184.35	158.30	0.63	180.46	154.97	0.57	170.99	146.83
0.9	0.5	0.37	318.17	273.21	0.36	305.04	261.94	**0.33**	**285.48**	**245.14**
	0.7	0.40	339.64	291.65	0.39	331.94	285.04	0.36	313.58	269.27
	0.9	0.43	367.05	315.19	0.43	369.16	316.10	0.41	353.96	303.94

4 Efficiency Comparisons

The percent relative efficiencies of the estimators $T_i(i = 1, 2)$ with respect to (i) $\bar{y}^*$, when there is no matching and (ii) Singh and Priyanka [1] estimator i. e. $\widehat{\bar{Y}} = \psi\bar{y}^* + (1 - \psi)\{\bar{y}_m + b_{yx}(\bar{x}_n - \bar{x}_m)\}$ have been obtained for different choices of correlations. For different choices of correlations, Tables 1 and 2 shows optimum values of μ_i ($i = 1, 2$) and percent relative efficiencies of $E_i^{(1)}$ and $E_i^{(2)}$ of $T_i(i = 1, 2)$ respectively with respect to $\bar{y}^*$ and $\widehat{\bar{Y}}$,

$$\text{Where } E_i^{(1)} = \frac{V(\bar{y}^*)}{M\left(T_i^{(o)}\right)_{opt}} \times 100 \text{ and } E_i^{(2)} = \frac{V(\widehat{\bar{Y}})}{M(T_i^{(o)})_{opt}} \times 100; (i = 1, 2). \quad (9)$$

Remark 4.1 The Tables 1 and 2 are constructed under the following considerations.

Table 2 Optimum values of μ_2 and percent relative efficiencies of T_2

ρ_{yz_1}		0.5			0.7			0.9		
ρ_{yz_2}	$\rho_{z_1z_2}$	$\mu_2^{(o)}$	$E_2^{(1)}$	$E_2^{(2)}$	$\mu_2^{(o)}$	$E_2^{(1)}$	$E_2^{(2)}$	$\mu_2^{(o)}$	$E_2^{(1)}$	$E_2^{(2)}$
ρ_{yx}	0.5									
0.5	0.5	0.62	121.63	122.43	0.60	119.66	120.45	0.58	117.78	118.56
	0.7	0.64	123.70	124.52	0.62	121.63	122.43	0.60	119.66	120.45
	0.9	0.67	125.88	126.71	0.64	123.70	124.52	0.62	121.63	122.43
0.7	0.5	0.51	164.59	165.68	0.50	161.30	162.37	0.49	158.24	159.29
	0.7	0.53	168.13	169.25	0.51	164.59	165.68	0.50	161.30	162.37
	0.9	0.54	171.97	173.11	0.53	168.13	169.25	0.51	164.59	165.68
0.9	0.5	0.23	297.26	299.23	0.23	290.92	292.85	0.23	284.96	286.84
	0.7	0.22	303.98	305.99	0.23	297.26	299.23	0.23	290.92	292.85
	0.9	0.20	311.03	313.09	0.22	303.98	305.99	0.23	297.26	299.23
ρ_{yx}	0.7									
0.5	0.5	0.71	128.14	110.03	0.67	126.24	108.40	0.65	124.39	106.81
	0.7	0.79	131.28	112.73	0.73	129.41	111.13	0.69	127.50	109.48
	0.9	1	133.33	114.49	0.85	132.39	113.68	0.77	130.67	112.21
0.7	0.5	0.55	172.78	148.36	0.53	169.48	145.53	0.52	166.40	142.89
	0.7	0.58	178.83	153.56	0.56	175.11	150.37	0.54	171.65	147.40
	0.9	0.63	185.68	159.44	0.60	181.47	155.83	0.57	177.56	152.47
0.9	0.5	0.21	306.76	263.42	0.22	300.98	258.45	0.23	295.44	253.70
	0.7	0.19	316.84	272.07	0.21	310.75	266.84	0.22	304.81	261.74
	0.9	**0.14**	**326.70**	**280.53**	0.17	320.90	275.56	0.20	314.80	270.32

i. N = 5000, n = 1200, k = 1.5 andW$_2$ = 0.25.

ii. "*" (in Tables 1 and 2) indicates no gain, i.e., PRE is less than 100 %.

iii. The expressions of $V(\bar{y}^*)$ and $V(\widehat{\bar{Y}})$ can be obtained from Singh and Priyanka [1].

5 Conclusions

From Tables 1 and 2, the following conclusions can be read-out from the present study.

(a) For fixed values of ρ_{yx}, ρ_{yz_1} and $\rho_{z_1z_2}$, the values of $\mu_1^{(o)}$, $\mu_2^{(o)}$ decrease while the values of $E_1^{(1)}\,E_1^{(2)}$, $E_2^{(1)}$ and $E_2^{(2)}$ are increasing with increasing choices of ρ_{yz_2}. These phenomenons are highly desirable, as they pay in terms of enhance precision of estimates and also reduce the cost of survey.

(b) Minimum value of $\mu_1^{(o)}$ and $\mu_2^{(o)}$ are 0.33 and 0.14 respectively which indicates that only 33 and 14 percent of the total sample size are to be replaced at the second (current) occasion for the corresponding choices of correlations, leading to appreciable reduction in cost.

Thus it is clear that our proposed estimators T_i ($i = 1, 2$) are preferable over the existing ones and if highly correlated auxiliary variables are used, only a smaller fraction of the sample on the second (current) occasion is required to be replaced by a fresh sample. These patterns are highly desirable as they reduce the cost of the survey.

References

1. Singh, G. N., Priyanka, K.: Effect of non-response on current occasion in search of good rotation patters on successive occasions. Stat. Transit. 8(2), 273–292 (2007)
2. Jessen, R. J.: Statistical investigation of a sample survey for obtaining farm facts. In: Iowa Agricultural Experiment Station Road Bulletin No. 304, Ames, pp. 1–104. (1942)
3. Singh, G. N., Karna, J. P.: Estimation of population mean under non-response in two-occasion rotations patterns. Commun. Stat. Theo-Meth. (2012, in Press). doi:10.1080/03610926.2012.698784
4. Singh, G.N., Homa, F.: Effective rotation patterns in successive sampling over two occasions. J. Stat. Theory Pract. 7(1), 146–155 (2013)
5. Hansen, M.H., Hurwitz, W.N.: The problem of the non-response in sample surveys. J. Am. Stat. Assoc. 41, 517–529 (1946)
6. Bahl, S., Tuteja, R.K.: Ratio and product type exponential estimator. Inf. Optim. Sci. 12, 159–163 (1991)

Modeling the Complex Dynamics of Epidemic Spread Under Allee Effect

Parimita Roy and Ranjit Kumar Upadhyay

Abstract An attempt has been made to investigate the dynamics of a diffusive epidemic model with strong Allee effect in the susceptible population and with an asymptotic transmission rate. We show the asymptotic stability of the endemic equilibria. Turing patterns selected by the reaction-diffusion system under zero flux boundary conditions have been explored. We have also studied the criteria for diffusion-driven instability caused by local random movements of both susceptible and infective subpopulations. Based on these results, we perform a series of numerical simulations and find that the model exhibits complex pattern replication: spots and spot–stripe mixture patterns. It was found that diffusion has appreciable influence on spatial spread of epidemics. Wave of chaos appears to be a dominant mode of disease dispersal.

Keywords Epidemic spread · Turing instability · Wave of chaos

1 Introduction

Since the pioneer work of Allee [1], there is an ongoing interest in the Allee effect on the dynamics of the population models [2]. The Allee effect is described by the positive relationship between any component of individual fitness and either numbers or density of conspecifics. The Allee effect can be caused by difficulties in finding mating partners at small densities, genetic inbreeding, demographic stochasticity or a reduction in cooperative interactions [2]. Here we show that the existence of a strong Allee effect (population decline at small densities) can lead to surprisingly rich dynamics in a basic epidemiological model. And the joint interplay of infectious diseases and Allee effects has been studied extensively in

P. Roy · R. K. Upadhyay (✉)
Department of Applied Mathematics, Indian School of Mines, Dhanbad, 826004 Jharkahnd, India
e-mail: ranjit.chaos@gmail.com

G. P. Biswas and S. Mukhopadhyay (eds.), *Recent Advances in Information Technology*, Advances in Intelligent Systems and Computing 266, DOI: 10.1007/978-81-322-1856-2_13, © Springer India 2014

epidemiology and lots of important phenomena have been observed. The aim of this paper is to explore the consequences of the Allee effect on the disease transmission as well as on the spatial spread. We investigate the wave of chaos phenomena and spatiotemporal pattern formation in a spatial SI model on spread of epidemic with asymptotic transmission rate.

The organization of the paper is as follows: Section 2 describes the development of the mathematical model. Section 3 is dedicated to stability analysis of the model system. In Sect. 4, the existence of the Turing instability analysis is discussed. Numerical results are presented in Sect. 5. Section 6 concludes the paper.

2 Development of the Model System

The following equations describe the epidemiological model:

$$
\begin{aligned}
dS/dt &= rS(1 - (S+I)/K)(1 - m/(S+I)) - \beta SI/(S+I+c) = F(S,I), \\
dI/dt &= \beta SI/(S+I+c) - aI = G(S,I),
\end{aligned}
\tag{1}
$$

where r is the intrinsic growth rate and K the carrying capacity, β denotes the contact rate, c represents half saturation constant and a denotes the disease-induced mortality. For the model (1), the basic reproduction is defined as $R_0 = \beta/a$. Also, assume that the susceptible (S) and infectious (I) population moves randomly, then a simple spatial model corresponding to Eq. (1) can be rewritten as:

$$
\begin{aligned}
\partial S/\partial t &= rS(1 - (S+I)/K)(1 - m/(S+I)) - R_0 aSI/(S+I+c) + D_S \nabla^2 S, \\
\partial I/\partial t &= R_0 aSI/(S+I+c) - aI + D_I \nabla^2 I,
\end{aligned}
\tag{2}
$$

where the nonnegative constants D_s and D_I are the diffusion coefficients of S and I, respectively. $\nabla^2 = \partial^2/\partial x^2 + \partial^2/\partial y^2$, is the usual Laplacian operator in two-dimensional space. The model is to be analyzed under the following non-zero initial condition: $S(p,0) \geq 0$, $I(p,0) \geq 0$, where $p = (x,y) \in \Omega = [0,L] \times [0,L]$, and zero-flux boundary conditions: $\partial S/\partial n = \partial I/\partial n = 0$, L denotes the size of the system in the direction of x and y; n is the outward unit normal vector on the boundary $\partial \Omega$.

3 Stability Analysis of the Model System

We perform the local stability analysis of both the systems without and with diffusion. This elucidates how dynamics of epidemic spread emerges from an interaction between local random movements of individuals of subpopulations and the dynamics of local kinetics.

3.1 Stability Analysis of the Temporal Model System

First, we will establish the local stability for the non-diffusive system (1).
The model system (1) has the following equilibrium points:

1. The disease free equilibrium point $E_1 = (K, 0)$ exists on the boundary of the first octant. For E_1, the eigenvalue are $((m/K) - 1)r$ and $a(KR_0/(c + K) - 1)$. Therefore the equilibrium point E_1 is locally asymptotically stable provided $(m/K) < 1$ and $KR_0/(c + K) < 1$. Also E_1 is unstable if $(m/K) > 1$ and $KR_0/(c + K) > 1$.

2. The disease free equilibrium point $E_2 = (m, 0)$ exists on the boundary of the first octant. For E_2, the eigenvalue are $(1 - m/K)r$ and $a(mR_0/(c + m) - 1)$. Therefore the equilibrium point E_2 is locally asymptotically stable provided $1 < m/K$ and $mR_0/(c + m) < 1$. Also E_1 is unstable if $1 > m/K$ and $mR_0/(c + m) > 1$.

3. The nontrivial equilibrium $E^*(S^*, I^*)$ exists if and only if there is a positive solution to the following set of equation:

$$g_1(S, I) = r(1 - (S^* + I)/K)(1 - m/(S^* + I^*)) - (R_0 a S^*)/(S^* + I^* + c) = 0,$$
(3a)

$$g_2(S, I) = (R_0 a S^*)/(S^* + I^* + c) - a = 0.$$
(3b)

Putting value of I^* from (3a) into (3b) we get,

$$g_1(S, I) = rS^*(1 - (R_0 S^* - c)/K)(1 - m/(R_0 S^* - c)) - a(R_0 S^* - S^* - c) = 0,$$
(3c)

S^* is the positive root of the cubic polynomial $p_1 S^{*^3} + q_1 S^{*^2} + r_1 S^* + s_1 = 0$ where $p_1 = rR_0^2$, $q_1 = (-(2c + K + m)r + aK(-1 + R_0))R_0$, $r_1 = (c + K)\ (c + m)r + acK(1 - 2R_0)$, $s_1 = ac^2K$. Since all the parameters are positive, we have $p_1 > 0$ and $s_1 > 0$. For if we have $R_0 > 1$ and $a > (2cr + Kr + mr)/(-K + KR_0)$ than we find that $q_1 > 0$ and $r_1 < 0$. Therefore, there is one positive root of the cubic polynomial. Also if $R_0 S^* > S^* + c \Rightarrow I^* > 0$. This completes the existence of $E^*(S^*, I^*)$.

The Jacobian of model around the endemic $E^*(S^*, I^*)$ is given by

$$V(E^*) = \begin{bmatrix} a_{11}^* & a_{12}^* \\ a_{21}^* & a_{22}^* \end{bmatrix}, \text{ where } a_{12}^* = S^*\left(\frac{mr}{(I^* + S^*)^2} - \frac{r}{K} - \frac{R_0(c + S^*)}{(c + I^* + S^*)^2}\right),$$

$$a_{21}^* = aI^*(c + I^*)R_0/(c + I^* + S^*)^2, \quad a_{22}^* = a\left(R_0 S^*(c + S^*)/(c + I^* + S^*)^2 - 1\right).$$

$$a_{11}^* = -\frac{I^*(c+I^*)R_0}{(c+I^*+S^*)^2} + r\left(\frac{-I^*+K+m}{K} - \frac{2S^*}{K} - \frac{I^*m}{(I^*+S^*)^2}\right), \quad (4)$$

Then the characteristic equation of $V(E^*)$ is given by $\lambda^2 + A_1\lambda + A_2 = 0$, where $A_1 = -(a_{11}+a_{22})$, $A_2 = (a_{11}a_{22} - a_{12}a_{21})$, Now according to Routh-Hurwitz criterion E^* is locally asymptotically stable if $A_1 > 0$ and $A_2 > 0$.

3.2 Stability Analysis of the Spatial Model System

In this section, we study the effect of diffusion on the model system about the endemic equilibrium point. Instability will occur due to diffusion when a parameter varies slowly in such a way that a stability condition is suddenly violated and it can bring about a situation wherein perturbation of a nonzero (finite) wavelength starts growing. In the two-dimensional case the model system can be written as in (2). To see the effect of diffusion we have considered the linearization of (2) at the positive equilibrium E^* and following the standard linear analysis of reaction diffusion system, we consider that (s, i) are small perturbation of (S, I) about the equilibrium point (S^*, I^*),

$$S(x,t) = S^* + s(x, y, t), I(x,t) = I^* + i(x, y, t). \quad (5)$$

The linearized system is given by

$$\frac{\partial s}{\partial t} = b_{11}s + b_{12}i + D_S\nabla^2 s, \quad \frac{\partial i}{\partial t} = b_{21}s + b_{22}i + D_I\nabla^2 i, \quad (6)$$

where, b_{ij} is same as a_{ij}^* for $i, j = 1, 2$ as given in Eq. (4). We write the solution of Eq. (6) in the form $s(x,y,t) \approx u \, \exp(\lambda_k t) \, \exp(i(k_x x + k_y y))$,

$$i(x,y,t) \approx v \, \exp(\lambda_k t) \, \exp(i(k_x x + k_y y)) \quad (7)$$

Substituting the above expressions for s, i into Eq. (6). The homogeneous equations in u and v have solutions if the determinant of the coefficient matrix is zero. i.e.,

$$|A_k - \lambda I| = 0, \quad (8)$$

where $A_k = A - k^2D$, I is the identity matrix and $D = diag(D_S, D_I)$ is the diffusion matrix and A is given by $A = \begin{pmatrix} b_{11} & b_{12} \\ b_{21} & b_{22} \end{pmatrix}$. Now Eq. (8) can be solved, yielding the so called characteristic polynomial

$$\lambda^2 + tr(A_k)\lambda + \det(A_k) = 0, \quad (9)$$

where $tr(A_k) = (D_S + D_I)k^2 - (b_{11} + b_{22})$, $k^2 = k_x^2 + k_y^2$,

$$\det(A_k) = D_S D_I k^4 - (D_I b_{11} + D_S b_{22})k^2 + b_{11}b_{22} - b_{12}b_{21},$$

Therefore, the solution of Eq. (9) is $\lambda = (-tr(A_k) \pm \sqrt{tr(A_k)^2 - 4\det(A_k)})/2$. By Routh–Hurwitz criterion, the roots of Eq. (9) are negative or has negative real parts if $tr(A_k) > 0$ and $\det(A_k) > 0$.

4 Turing Instability

Turing instability means that a steady state is asymptotically stable for the non-spatial model system (1) but is unstable with respect to solutions of spatial model system (2). In fact, the Turing instability sets in when at least one of solutions of the Eq. (9) crosses the imaginary axis. In other words, the spatially homogeneous steady state will become unstable due to heterogeneous perturbation when at least one solution of the Eq. (9) is positive. For this reason, at least one out of the following two inequalities is violated

$$tr(A_K) = b_{11} + b_{22} - (D_S + D_I)k^2 < 0 \tag{10}$$

$$\det(A_k) = D_S D_I k^4 - (D_S b_{22} + D_I b_{11})k^2 + b_{11}b_{22} - b_{12}b_{21} > 0 \tag{11}$$

Since perturbation of zero wave number are stable by definition (due to the stability for non-spatial steady state), $tr(A) = a_{11} + a_{22} < 0$ and $\det(A) = a_{11}a_{22} - a_{12}a_{21} > 0$ are satisfied. It is seen from these facts that the condition (10) always holds. Hence we are left only one for instability condition (11), i.e., $H(k^2) = D_S D_I (k^2)^2 - (D_S b_{22} + D_I b_{11})k^2 + b_{11}b_{22} - b_{12}b_{21} < 0$ for some k.

For $H(k^2)$ be negative for some non-vanishing k, the minimum of $H(k^2)$ denoted by $H_{\min}$ must be negative. The function $H(k^2)$ has minimum $H_{\min}$ for some value k_T^2 of k^2 at $k_T^2 = D_S b_{22} + D_I b_{11}/(2D_S D_I)$.

Then the condition that $H(k_T^2) < 0$ gives $(D_S b_{22} + D_I b_{11})^2 > 4D_S D_I (b_{11}b_{22} - b_{12}b_{21})$.

For the parameters value $r = 2.1, K = 400, c = 10, \beta = 5.1, a = 0.86, m = 0.15$. The equilibrium $E^* = (19.022871, 39.547943)$ is locally asymptotically stable $(A_1 = 0.308434, A_2 = 0.58053)$.

Now we, plot Fig. 1, the dispersion relation corresponding to the same parameter value $r = 2.1, a = 0.86, K = 400, c = 10, m = 0.15, \beta = 3.1, D_S = 0.01, D_I = 5$, it is easy to see that the real part of λ is positive for some region $0.802858 = k_1 < k < k_2 = 4.24416$. That is to say, with the fixed parameter values, if $0.802858 = k_1 < k < k_2 = 4.24416$, the solutions of model system (2) are unstable.

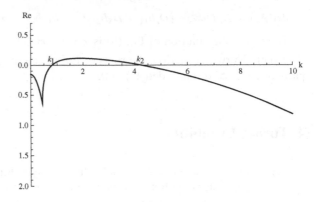

Fig. 1 Variation of dispersion relation of model (2) with $\beta = 3.1$, the Turing intability occurs

5 Simulations Results

The dynamics of the model system (2) is studied with the help of numerical simulation, both in one and two dimensions. We use the standard five-point approximation for the 2D Laplacian with the Neumann (zero-flux) boundary conditions. In this case, we see that the small random perturbation of the stationary solution S^* *and* I^* leads to the formation of a strongly irregular transient pattern in the domain when the parameter values are in the domain of Turing space. In this subsection, the plots (space vs. population densities) are obtained to study the spatial dynamics of the model systems. In one dimensional case, we assume domain size 7000. From a realistic biological point of view, we consider a non monotonic form of initial condition which determine the initial spatial distribution of the species in the real community as $S(x,0) = S^* + \varepsilon(x - x_1)(x - x_2), I(x,0) = I^*$, where (S^*, I^*) the non-trivial is state for co-existence of susceptible and infective and $\varepsilon = 10^{-8}$, $x_1 = 1200, x_2 = 2800$ is the parameter affecting the system dynamics. The dynamics of the susceptible and infective is observed at the parameter values $r = 2.1, \beta = 5.1, K = 400, a = 0.86, m = 0.15, D_S = 0.001$, and at time level $t = 500$ for two different diffusivity constant i.e. $D_I = 5, 10$ as shown in Fig. 2. The ranges of values of the parameters are chosen on the basis of the values reported in Jorgensen [4] and the previous study [3]. As we increase the diffusivity constant the size of the domain occupied by the irregular chaotic patterns slowly grows with time in both directions displacing the regular pattern (characterized by a stable limit cycle in the phase plane of the system) with chaotic dynamics. This phenomenon has been termed as "Wave of Chaos" (WoC) [5].

5.1 Disease Spread with Varying D_S and Fixed D_I

In this case, we consider spatiotemporal dynamics of model (2) with fixed parameters $r = 2.1, K = 400, \beta = 5.1, a = 0.86, c = 10, m = 0.15, D_I = 5.$ and

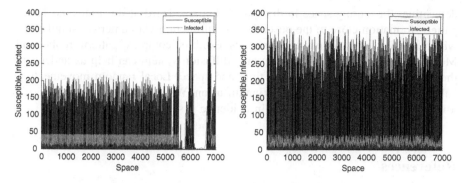

Fig. 2 Space series generated at different time $t = 500$ for two values of $D_I = 5, 10$ showing the effect of diffusivity constant D_I on dynamics of the model system (2)

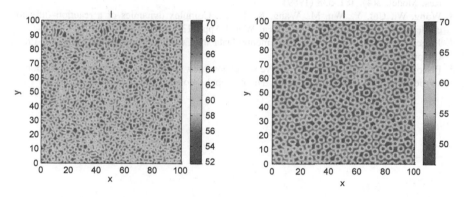

Fig. 3 Spatiotemporal dynamics of varying D_S at $t = 200$ days **a** $D_S = 0.015$; **b** $D_S = 0.05$. This shows that infection increases with increase of D_S

varying parameter D_s. Parameter values are chosen from Turing Space. Turing patterns are shown in Fig. 3. Clusters of higher densities of infective which are distributed over the whole domain are discernible in Fig. 3. When D_s varies from 0.015 to 0.05, a sequence "small spot → large-spot" is observed.

6 Discussions and Conclusions

In this paper, we analyze a reaction- diffusion model with the strong Allee effect under the zero-flux boundary conditions in both one and two dimensions. This study lies in three-folds. First, presents the stability of the equilibrium with the strong Allee effect, which indicates that the dynamics of the model with Allee effect are rich and complex. Second, gives an analysis of Turing instability, which determines the Turing space in the spatial domain. Third, it illustrates the Turing pattern formation close to the onset Turing bifurcation via numerical simulations, which shows that the model dynamics exhibits complex pattern replication. Modeling the epidemics using reaction—diffusion system can help us understand the distribution of disease in both time and space. Local random movements of subpopulations are modeled by Fickian diffusion. We believe that wave of chaos is a mechanism for disease dispersal in epidemic systems.

References

1. Allee, W.C.: Animal Aggregations: A Study in General Sociology. AMS Press, New York (1978)
2. Dennis, B.: Allee effects: population growth, critical density, and the chance of extinction. Nat. Res. Model. **3**(4), 481–538 (1989)
3. Wang, W., Cai, Y., Wu, M., Wang, K., Li, Z.: Complex dynamics of a reaction–diffusion epidemic model. Nonlinear Anal.: Real World Appl. **13**(5), 2240–2258 (2012)
4. Jorgensen, SE.: Handbook of Environmental Data and Ecological parameters. Pergamon Press, Oxford (1979)
5. Petrovskii, S.V., Malchow, H.: Wave of chaos: new mechanism of pattern formation in spatiotemporal population dynamics. Theor. Popul. Biol. **59**, 157–174 (2001)

Rainfall Prediction Using k-NN Based Similarity Measure

Arpita Sharma and Mahua Bose

Abstract Time Series Analysis is helpful in understanding past and current pattern of the phenomenon and provides important clue for predicting future trends. Weather forecasting is a matter of great importance in the area of Time Series Analysis. The purpose of this paper is to predict the monthly rainfall based on the historical dataset using pattern similarity based models together with k-NN technique and to compare the estimated values with the actual observations. We have applied a recently proposed approach Approximation and Prediction of Stock Time-series data (APST) for forecasting rainfall. Further we have also suggested two variations of APST. We have also compared the results of similarity based methods with Autoregressive Model. Suggested techniques yield better results than original APST and AR Models.

Keywords APST · Auto regression · Forecast · Rainfall · Pattern similarity · k-NN · Time series

1 Introduction

Time series is a sequence of quantitative observations measured successively at uniform time intervals. Time series data have a natural temporal ordering of the observations. Though the primary goal of time series analysis is forecasting, it can also be used for clustering, classification, anomaly detection and query by content. It has application in Economics, statistics, agriculture, seismology, meteorology, geophysics, transportation, employment sector, tourist trips etc.

A. Sharma (✉) · M. Bose
Delhi University, New Delhi, India
e-mail: asharma@ddu.du.ac.in

M. Bose
e-mail: e_cithi@yahoo.com

G. P. Biswas and S. Mukhopadhyay (eds.), *Recent Advances in Information Technology*, 125
Advances in Intelligent Systems and Computing 266, DOI: 10.1007/978-81-322-1856-2_14,
© Springer India 2014

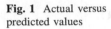

Fig. 1 Actual versus
predicted values

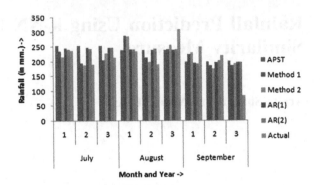

Rainfall in India is primarily caused by South–West monsoon. Prediction of the spatial and temporal variability in rainfall is a major problem in an agricultural country like India. Any fluctuations in the monsoon not only affect the economy of an area but also the vegetation and soil characteristics of the area. Advance knowledge about rainfall helps in decision making. It has great impact on flood control.

In recent years notable research works have been done in the areas of climatology and hydrology especially in the field of rainfall prediction using various statistical and soft computing techniques. Studies that use various regression based techniques namely AR, ARMA, ARIMA have been successfully attempted in the past [1–3]. Bayesian hierarchical model [4], Non homogeneous Hidden Markov Model and statistical coherence analysis [5], linear regression based methods [6] were some of the other approaches. Notable works based on K-Nearest Neighbour methods [7, 8], Artificial Neural Network [9–14], Association rule mining [15] and fuzzy logic [16–18] were also proposed. Soft computing techniques were also attempted successfully [19, 20].

Recently Approximation and Prediction of Stock Time-series data (APST) approach was presented [21], which is a two step approach to predict the direction of change of stock price indices. In this paper we have (i) Applied APST approach for rainfall forecasting (ii) Proposed two variations of this technique (iii) Performed experiments on a dataset which has rainfall data for 150 years and compared the results obtained for original APST and our proposed variations with the AR(1) and AR(2) Model. The results so obtained are shown in Fig. 1.

This paper is organized as follows: Sect. 2 presents the dataset used for validation of our algorithm and in Sect. 3 tests are conducted to determine stationarity of data. Procedural details are presented in Sect. 4. Experimental results are discussed in Sect. 5. Section 6 summarizes our conclusion.

2 Data

We have collected about 150 year's rainfall data from the website of Indian Institute of Tropical Meteorology [22]. In our experiment we shall use rainfall data of East Peninsular India from 1863 to 1997 (July–Sept.). Data for the year 1998–2006 will be used for error estimation purpose.

Fig. 2 Yearly rainfall pattern

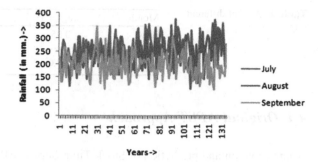

3 Test for Stationarity in Data Series

Before the experiment, we have checked for stationarity of data series. In a stationary process, parameters such as the mean and variance do not change over time and do not follow any trend. Yearly rainfall data (Fig. 2) shows up and down movement with little or no evidence of a trend.

We have performed Dickey–Fuller test [23] to determine whether a unit root is present in autoregressive model. A simple AR(1) model is defined as : $y_t = \rho\, y_{t-1} + e_t$, where y_t is the variable of interest, t is the time index, ρ is a coefficient, and e_t is the error term. If unit root is present ($\rho = 1$), the model would be non-stationary. The value of ρ obtained using above equation for three data sets are 0.043, 0.05, −0.023 indicating stationarity in the data. In addition to that, we have calculated Autocorrelation function (ACF) and Results are shown in Table 1.

4 Forecast Models

This section presents the base algorithm APST and suggested variations of APST. These algorithms are based on pattern similarity measure rather than traditional statistical techniques. Let, a time series be $f = (f_1, f_2, \ldots\ldots, f_n)$, where time $t = 1, 2, \ldots\ldots n$. If it is required to forecast for next m periods, i.e., $f_{n+1}, \ldots\ldots f_{n+m}$, continuation is to be based on past values, i.e., $f_{n-m+1}, \ldots\ldots, f_n$. The purpose of this method is to identify the most similar series to the immediate past series.

In Auto regression based forecasting models such as AR, ARMA, etc., to get a forecast for next period, data for the preceding (depends upon the order of model) periods are required. Here, forecast for a single value is obtained at a time and to get a forecast for subsequent years, predicted values from prior steps should be incorporated into the model. So, they are iterative in nature. But similarity based methods, give direct forecast for the n periods ahead.

Table 1 ACF at different Lag

Month	Lag = 1	Lag = 2	Lag = 3
July	0.05	0.09	0.12
Aug	0.06	0.09	−0.01
Sept	−0.03	0.06	−0.02

4.1 Original APST Approach

Approximation and prediction of Stock Time-Series (APST) algorithm is based on the idea of label based forecasting technique (LBF) [24]. It is a pattern sequence based prediction technique consisting of two phases: (1) Generation of sequence of approximated values from the past data using multiscale segment mean approach [25]. (2) Forecasting using these approximated values. Original APST algorithm is applied for daily stock exchange forecast. We have applied this algorithm for monthly rainfall prediction based on past dataset.

Data Approximation: Data is divided into K groups containing 27 consecutive elements and each t sub segment contains three consecutive elements from each K groups. Thus within a group, data is organized into a tree structure where each node has three children. Total number of levels in the tree is $\log_t K$. Data approximation of input time series is obtained by computing segment means from lower level to upper level in the tree structure. Three adjacent segment of lower level constitute a segment in upper level. Leaf nodes contain original data values and each parent node is mean value of its children. Here, Euclidean distance has been used.

Data Prediction: The original Algorithm [21] for prediction is given below. Here, $Ap = [Ap_{10}, Ap_{20}....., Aps_{n0}]$ is the set of approximated values obtained by step1.

```
begin
Ap= [Ap₁₀, Ap₂₀ . . . . . . . .,Apsₙ₀]
         PS=<Apₙ₋w, Apₙ₋w₋₁,. . . . . . ., Apₙ>
         NN=find the nearest neighbours for PS in AP<nn₁,nn₂,..............,nnₖ>
         for each nnᵢ ∈ NN do
         Eᵢ =Extract sequence < eᵢ₁, eᵢ₂,.....,eᵢₘ> of m elements next to nnᵢ
         end for
         for each j=1 to m do
                 for each element Eᵢ ∈ E do
                 P'[j]=p'[j]+ eᵢⱼ
                 end for
         end for
end
```

The above was applied by taking m = 3 (m nearest values are taken) but it can be any number to minimize error. In future, error can be further reduced by adjusting number of neighbours. It also depends on the volume of dataset. For large dataset we can take the large sequences and large number of neighbouring values.

4.2 Modified Approach

In our modified algorithm (henceforth called as method 1), to get a forecast for the next N periods, calculations are based on average of previous N consecutive values. As, in this experiment, forecast for next three periods are to be generated (N = 3), average of three consecutive values are calculated. So, we have divided data points into groups (each containing consecutive three data points) as in APST. Here, we have 45 groups. But no further grouping is done. Each group is identified by a number. For example group-id one contains 1–3, two contains 4–6. Distance is calculated from group average of last group to all other groups. First K shortest distances are taken as neighbours and corresponding group-ids are retrieved. Data points belonging to those groups are extracted and their mean is taken year wise. These are the next 3 year's forecast. This experiment is carried out by taking K = 3 and it should be selected carefully to get accurate result.

For more than three periods forecast, say for forecast for next four periods, mean of four consecutive data values should be computed and so on. Algorithm of Method 1 is given below.

Input: an array of n group averages and their corresponding group-ids
Output: forecast (for a particular month) for Next three years

```
begin
for all averages i=1 to n-1
        diff [i] = |avg[n]-avg[i]|   // it stores distance between two averages
end for
fori=1 to (n-1)-1
for j=i+1 to n-1                     // distances are sorted in asc. order along with
group-id
        if(diff[i]>diff[j])
        Interchange elements in diff and group-id
        endif
        end for
end for
        NN=find the nearest neighbours for diff  <dn₁,dn₂,..............dnₖ>
        for each group-id corresponding to dnᵢ ∈  NN do
            Eᵢ =Extract sequence < eᵢ₁, eᵢ₂, eᵢ₃ > of 3 elements from the corre-
        sponding group
        end for
        for each sequence j=1 to 3 do          // Here, no. of neighbor=3
                find average of corresponding elements
        end for
end
```

We have proposed another modification (henceforth called as method 2) in which instead of taking average of corresponding elements for a particular year, a weighted average scheme has been proposed. Here, elements in each sequence are multiplied by the corresponding group distance stored in diff array. Sum of the corresponding elements in three sequences are divided by the sum of differences.

4.3 Autoregressive Model

AR model of order p is defined as: $y_t = c + \varphi_1 y_{t-1} + \varphi_2 y_{t-2} + \ldots\ldots\ldots\ldots\ldots\ldots + \varphi_p y_{t-p} + e_t$, where c is the constant, φ terms are coefficients and e_t is the error.

For an AR(1) process only the previous term and the noise term contribute to the output. For an AR(2) process, the previous two terms and the noise term contribute to the output.

We want to forecast y_{t+m} for 1.........m period, where, m = 1 means one-step ahead and m > 1 means multi-step ahead forecasting. Now, the above equation gives forecast for next period or one-step ahead. So, it is the prediction for the first unobserved period. The next's next period, for which data is not yet available, the predicted value calculated from the previous forecasting step is used. Thus, for multi-step (k) ahead forecasting, terms that have not happened at t are replaced by their forecasts at t.

5 Experimental Results and Discussion

Experiments are carried on three rainfall data set (July–Sept.). Earlier we have checked that the data are stationary. We have 135 data points (for each of 3 months) for experiment and 9 for testing. For APST, at top level, we have five values (135/27). Each of them has three children. There are total nine nodes in the next level. Thus, total 27 leaf nodes are there for each tree. Original APST algorithm gives good result for next year's forecast but failed to predict properly for subsequent years. For example, in case of July data, if we consider 3-NN and set size of predicted values = 9, RMSE = 66. For size of predicted values = 3, RMSE = 45. For our modified algorithms result are noted for next 3 years and better results are obtained. RMSE, MAE and MER value for Method 1 on August data is higher than that of APST but on other dataset it performs better than APST. Method 2, i.e., weighted average technique is best in all the cases. For August and September dataset, APST and AR models show almost similar results. Results of our experiment are shown in Table 2.

Prediction for the third year in September is far from actual observation in all of the methods. AR models proved good for next year's forecast only. Using AR(1) and AR(2), forecast for next period (year 1) in July and August are better than that of other three methods. But these methods failed to predict accurately for the other 2 years and Mean Square Error (MSE) is increased.

Table 2 Performance of different Methods

Month	Error	APST	Method 1	Method 2	AR(1)	AR(2)
July	MAE	41	4.26	12.59	33.55	29.82
	MER	19.16	1.99	5.87	15.68	13.93
	RMSE	45.28	4.95	15.05	39.08	36.05
August	MAE	41.58	48.3	38.42	42.32	43.17
	MER	16.94	19.68	15.65	17.24	17.59
	RMSE	50.13	51.84	43.77	49.73	50.55
September	MAE	61.3	52.12	56.67	62.6	61.44
	MER	33.16	28.19	30.65	33.86	33.26
	RMSE	73.2	62.25	67.83	72.99	72.96

6 Conclusion

Rainfall is a phenomenon which hardly follows any rule or predefined sequence. So it is not fair to forecast rainfall for next 5–10 years in advance using similarity based techniques. As rainfall depends on several atmospheric factors, these factors should be taken into consideration first. These methods are suitable for short term forecasting. For the next year's forecast these methods perform well. In this paper we have proposed two modifications of the APST algorithm and applied these on the chosen data set. We have also applied the original APST algorithm and Autoregressive model on the same dataset for comparison. In general proposed Method 2 outperforms APST and other method. In future, increased accuracy in prediction can be achieved by adjusting the number of elements in groups and the number of neighbours.

References

1. Burlando, P., Rosso, R., Cadavid, L.G., Salas, J.D.: Forecasting of short-term rainfall using ARMA model. J. Hydrol. **144**(1–4), 193–211 (1993)
2. Valipour, M.: Number of required observation data for rainfall forecasting according to climate conditions. Am. J. Sci. Res. **74**, 79–86 (2012)
3. KhadarBabu, S.K., Karthikeyan, K., Ramanaiah, M.V., Ramanah, D.: Prediction of rainfall-flow time series using auto-regressive models. Adv. Appl. Sci. Res. **2**(2), 128–133 (2011)
4. Henley, B.J., Thyer, M.A., Kuczera, G.: Seasonal stochastic rain fall modelling using climate indices: a Bayesian hierarchical model. In: International Congress on Modelling and Simulation, pp. 1575–1581 (2007)
5. Verbist, K., Robertson, A.W., Cornelis, W.M., Gabriels, D.: Seasonal predictability in daily rainfall characteristics in central northern Chile for dry-land management. J. Appl. Meteorol. Climatol. **49**, 1938–1955 (2010)
6. Sharma, A. Bose, M.: Seasonality and rainfall prediction. In: Seventh International Conference on Data Mining and Warehousing (ICDMW), pp. 145–150, Bangalore (2013)
7. Wu, A.: A novel artificial neural network ensemble model based on K-nearest neighbor nonparametric estimation of regression function and its application for rainfall forecasting. Comput. Sci. Optim. **2**, 44–48 (2009)

8. Jan, Z., Abrar, M., Bashir, S., Mirza, A.M.: Seasonal to Inter Annual Prediction Using Data Mining K-NN Technique. CCIs. vol. 20, pp. 40–51, Springer, New York (2008)
9. Olaiya, F., Adeyemo, A.B.: Application of Data Mining Techniques in Weather Prediction and Climate Change Studies. Int. J. Inf. Eng. Electr. Bus. **1**, 51–59 (2012)
10. Wu, C.L., Chau, K.W., Fan, C.: Prediction of rainfall time series using modular artificial neural networks coupled with data-preprocessing techniques. J. Hydrol. **389**(1–2), 146–167 (2010)
11. Luc, K.C., Ball, J.E., Sharma, A.: An application of artificial neural network for rainfall forecasting. Math. Comput. Modell. **33**(6–7), 683–693 (2001)
12. Ramirez, M.C.V., Velho, H.F.D.C., Ferreira, N.J.: Artificial neural network technique for rainfall forecasting applied to Sao Paulo region. J. Hydrol. **301**(1–4), 146–162 (2005)
13. Fallah-Ghalhary, G.A., Mousavi-Baygi, M., Habibi-Nokhandan, M.: Seasonal rainfall forecasting using artificial neural network. J. Appl. Sci. **9**(6), 1098–1105 (2009)
14. Hung, N.Q., Babel, M.S., Weesakul, S., Tripathi, N.K.: An artificial neural network model for rainfall forecasting in Bangkok, Thailand. Hydrol. Earth Syst. Sci. Dis. **5**(1), 183–218 (2008)
15. Sivaramakrishnan, T.R., Meganathan, S.: Association rule mining and classifier approach For quantitative spot rainfall prediction. J. Theor. Appl. Inf. Technol. **34**(2), 173–177 (2011)
16. Pao-Shan, Y., Shien Tsang, C., Che-Chuan, W., Shu-Chen, L.: Comparison of grey and phase-space rainfall forecasting models using a fuzzy decision method. Hydrol. Sci. **49**(4), 655–672 (2004)
17. Suwardi, A., Takenori, K., Shuhei, K.: Neuro-fuzzy approaches for modeling the wet season tropical rainfall. Agric Inf. Res. **15**(3), 331–341 (2006)
18. Fall, G.A., Mousavi-Ba, M., HabibiNok, M.: Annual rainfall forecasting by using Mamdani fuzzy inference system. Res. J. Environ. Sci. **3**(4), 400–413 (2009)
19. Wu, C.L., Chau, K.W.: Prediction of rainfall time series using modular soft computing methods. Eng. Appl. Artif. Intell. **26**(3), 997–1007 (2013)
20. Wong, K.W., Wong, P.M., Gedeon, T.D., Fung, C.C.: Rainfall prediction model using soft computing technique. Soft Comput.-SOCO. **7**(6), 434–438 (2003)
21. Vishwanath, R.H., Leena, S.V., Srikantaiah, K.C., Shreekrishna Kumar, K. Deepa Shenoy, P., Venugopal, K.R., Patnaik, L.M.: APST: Approximation and prediction of stock time-series data using pattern sequence. In: Venugopal, K.R., Deepa Shenoy, P., Patnaik, L.M (eds.) Seventh International Conference on Data Mining and Warehousing (ICDMW), pp. 151–160. Elsevier Publications, Bangalore (2013). ISBN: 978-93-5107-105-1
22. http://www.tropmet.res.in
23. Dickey, D.A., Fuller, W.A.: Distribution of the estimators for autoregressive time series with a unit root. J. Am. Stat. Assoc. **74**(366), 427–431 (1979) JSTOR 2286348
24. Martínez-Alvarez, F., Troncoso, A., Riquelme, J.C., Aguilar Ruiz, J.S.: LBF: a label-based forecasting algorithm and its application to electricity price time series. In: IEEE International Conference on Data Mining (2008)
25. Lian, X., Chen, L., Lian, X., Xu Yu, Jeffrey, Han, J., Ma, J.: Multiscale representations for fast pattern matching in stream time series. IEEE Trans. Knowl. Data Eng. **21**(4), 568–581 (2009)

Dimension Reduction of Gene Expression Data for Designing Optimized Rule Base Classifier

Amit Paul, Jaya Sil and Chitrangada Das Mukhopadhyay

Abstract The paper highlights the need of dimension reduction of voluminous gene expression microarray data for developing a robust classifier to predict patients with cancerous genes. The proposed algorithm builds a fuzzy rule based classifier with optimized rule set without much sacrificing classification accuracy. The gene expression matrix is first discretized using linguistic values. The importance factor of each gene is then evaluated representing the degree of presence of a unique linguistic value of the gene both in disease and nondisease classes. Initial fuzzy rule base consists higher ranking genes and gradually other genes are included in the rule base in order to achieve maximum classification accuracy. Thus optimum rule set is built with important genes for classification of test data set. The methodology proposed here has been successfully demonstrated for the lung cancer classification problem, which includes 97 smokers with lung cancer and 90 without lung cancer gene expression data. The results are promising even though maximum number of genes are removed from the original data.

Keywords Fuzzy rule base · Linguistic variable · Fuzzy importance

A. Paul (✉)
Computer Science and Engineering, St. Thomas College of Engineering and Technology, Khidirpore, India
e-mail: amitpaul83@gmail.com

J. Sil
Computer Science and Technology, Bengal Engineering and Science University, Shibpur, India
e-mail: js@cs.becs.ac.in

C. Das Mukhopadhyay
Health Care Science and Technology, Bengal Engineering and Science University, Shibpur, India
e-mail: chitrangadadas@yahoo.com

G. P. Biswas and S. Mukhopadhyay (eds.), *Recent Advances in Information Technology*, 133
Advances in Intelligent Systems and Computing 266, DOI: 10.1007/978-81-322-1856-2_15,
© Springer India 2014

1 Introduction

Machine learning and data mining techniques have been successfully applied [1–4] for long in biomedical data analysis to extract knowledge [5–7]. Dimensionality reduction is very much relevant in bio-informatics research, particularly in the context of microarray data, characterized by relatively few samples in a high-dimensional gene (feature) space. Irrelevant genes (features) lead to insufficient classification accuracy and add extra difficulties in finding potentially useful knowledge [8, 9]. Gene selection becoming one of the main sub-fields in bio-informatics data mining [10–12]. In the context of classification, the main goal of gene selection is to search for an optimal gene subset that lead to improved classification performance. During the past decades, extensive research has been conducted from multidisciplinary fields including statistics, pattern recognition, machine learning and data mining [10, 13, 14].

Continuous valued attributes generate large rule set and most classifiers with such huge number of rules are unstable even for slight change of training data set. Fuzzy rule based systems have been successfully applied to various application areas such as control, decision making, classification and many more [15, 16]. Some of the machine learning based works which generate fuzzy if-then rules are demonstrated in [17–19]. A version of fuzzy-ID3 algorithm which induces fuzzy decision trees is proposed in [17]. While the main objective in designing fuzzy rule-based systems is maximization of performance, comprehensibility of the system has also been taken into account in some recent studies [20–26]. The comprehensibility of fuzzy rule-based systems is related to linguistic interpretability of each fuzzy set in the rule, separation of neighboring fuzzy sets and the number of fuzzy sets for each linguistic variable. Simplicity of fuzzy rule based systems (e.g. the number of input variables, the number of fuzzy if-then rules) and fuzzy reasoning power are the other important factors that determine comprehensibility of the system.

In this paper, fuzzy if-then rules are optimized using selected genes to design a robust comprehensible fuzzy rule base for gene pattern classification problem with continuous attribute values. In the proposed method, first continuous value of microarray gene expression data are discretized using linguistic values determined by the variance of each sample corresponding to each gene. The linguistic values are relevant for discriminating alternative phenotypes and represent activation level of the gene. With the linguistic value, gene expression matrix is rebuilt. As a next step, fuzzy logic has been applied on the linguistic gene expression matrix to evaluate importance of each gene. Based on the importance factor, genes are ranked representing their significance in sample classification. Finally, fuzzy rules are framed with higher ranking genes and gradually next higher ranking genes are included in the fuzzy rule base with an objective to achieve maximum classification accuracy. Thus optimized rule set is obtained to classify patients with cancerous genes using test microarray gene expression data set.

The rest of the paper is organized as follows. The proposed method is described in Sect. 2. Experimental results are presented in Sect. 3. Concluding remarks and suggestions for future work are given in Sect. 4.

2 Methodology

Classification is a supervised learning technique that aims at exploring proper class of given objects based on the similarity among the objects. In disease classification problem, it has been noticed that although there are thousands of genes for each observation, a few underlying genes may account for much of the data variation. For instance, many of the genes may not be relevant to the tumour metabolic process, so they are potentially noise features. Removing such noise features may help to obtain higher classification accuracy (better diagnosis), resulting identification of marker genes. Selected genes increase classification accuracy with less number of comparisons which is ultimate goal of gene research.

2.1 Optimized Fuzzy Rule Generation (OFRG) Algorithm

Feature selection process refers to choosing a subset of attributes from the set of original attributes. The purpose of feature selection is to identify the significant features, eliminate the irrelevant features and build a robust learning model.

The proposed gene selection algorithm evaluates fuzzy importance factor of each gene that signifies relevance of the gene in classifying the diseased patients using microarray gene expression data.

2.2 Terminologies for the Proposed OFRG Algorithm

Linguistic Gene Expression $\left(LGE_{(i,j)}\right)$: To obtain linguistic gene expression value, first mean of each gene, say i for all samples are calculated, called *gean_mean$_{(i)}$*. Difference between the jth sample value of ith gene and mean of the sample values of the corresponding gene (*gean_mean$_{(i)}$*) as defined in (1) is used to assign the linguistic value at ($LGE_{(i,j)\text{th}}$) position of the matrix. Thus the discretized gene expression matrix is constructed with the *Linguistic gene expression* value, which represent activity level of gene i for sample j.

$$LGE_{(i,j)} = gene\_expression_{(i,j)} - gene\_mean_{(i)} \tag{1}$$

Gene Importance Factor$_{(i)}$: Determines maximum number of unique *Linguistic Gene Expression* value for ith gene appear in the disease classes.

Unconstructive Impact Factor$_{(i)}$: Determines number of unique *Linguistic Gene Expression* value present in *gean$_{(i)}$* for non-cancer classes. The *Unconstructive Important Factor* of *gean$_{(i)}$* measures its contributions towards calculating the *Gene Importance Factor*, which leads to misclassification of data.

Fuzzy Importance Factor (FIF$_{(i)}$): Distribution of gene expression data in two-dimension space reveals the fact that several genes are highly important compare to others in context of disease classification. Maximum number of samples with same *Linguistic Gene Expression* value in *i*th *gene* is used to evaluate importance of that gene. However, for *gene$_{(i)}$* same linguistic value may appear in different non-cancer classes too (*Unconstructive Impact Factor*) which affects classification accuracy. Therefore, *Fuzzy Importance Factor* is proposed to isolate the *Unconstructive Impact Factor* from *Gene Importance Factor* using relation (2).

$$FIf_{(i)} = \frac{k - m}{n} \quad if \ \ k > m$$
$$= 0 \qquad otherwise \tag{2}$$

where k, m and n represent *Gene Importance Factor, Unconstructive Impact Factor* of *i*th gene and total number of samples in the gene expression matrix respectively. *FIF$_{(i)}$* is zero when *i*th *Gene Importance Factor* is less/equal to the *Unconstructive Impact Factor* signifying that if *Unconstructive Impact Factor* is higher than the *Gene Impact Factor* then that gene has no effect to identify the proper class.

Algorithm 1: Linguistic value Generation Algorithm

Input   : Gene matrix *gene* having m and n gene and sample respectively
Output: Linguistic gene matrix *LGE*.
$mean\_g(i) := \text{mean}(gene(i)) \ \ \forall \ \ i \in \{1, 2 \ldots, m\}$
$nor\_v(i, j) := \lfloor gene(i, j) - mean\_g(i) \rfloor \ \forall i \in \{1, 2 \ldots, m\}; \ \forall j \in \{1, 2 \ldots, n\}$
if $nor\_v(i, j) < -10$ **then**
| $LGE(i, j) :=$very_low
else if $-10 \leq nor\_v(i, j) < -5$ **then**
| $LGE(i, j) :=$ low
else if $-5 \leq nor\_v(i, j) < 5$ **then**
| $LGE(i, j) :=$ moderate
else if $5 \leq nor\_v(i, j) < 10$ **then**
| $LGE(i, j) :=$ high
else
| $LGE(i, j) :=$ very_high

As the first step of the proposed method linguistic gene expression values are generated by applying Algorithm (1). Five different linguistic values are assigned to the samples of each gene by evaluating Eq. (1). Algorithm (2) is applied on discretized gene expression matrix to calculate fuzzy importance factor of each gene. Genes are clustered based on their similarity determined by *Fuzzy Importance Factor*. Finally, optimized fuzzy rule base is built with linguistic value of genes having maximum fuzzy importance factor as described in Algorithm 3.

3 Results and Comparisons

3.1 Data Set

Genes selected from microarray data sets of smokers with lung cancer and without lung cancer are compared. Database record series GSE4115 [27, 28] consisting of 22215 genes and each gene having 192 samples are considered for validating the scheme using Affymetrix genechip U133A.

3.2 Performance Measurements

OFRG algorithm has been applied on microarray data set to classify the samples by developing a fuzzy rule base classifier. Initial fuzzy rule base is built based on highest fuzzy importance factor and corresponding linguistic value are assigned for a specific disease class. Subsequently genes are included in the fuzzy rule base depending on their importance with an objective to improve classification accuracy.

Algorithm 2: Gene's Fuzzy Importance Algorithm

Input : Linguistic gene matrix LGE having m and n gene and sample respectively

Output: Fuzzy Importance factor matrix FIf.

$t(i) :=$ [Size(find($LGE(i)$, very_low)), Size(find($LGE(i)$, low)), Size(find($LGE(i)$, moderate)), Size(find($LGE(i)$, high)), Size(find($LGE(i)$, very_high))];

$GI := \max(t(i))$

$p := $ find($t(i)$, $\max(t(i))$);

$UC := $ Size(find($NLGE(i)$, $SLGE$))

$temp := GI - UC$

if $temp > 0$ then

$\quad | \quad FIf := \frac{temp}{n}$

else

$\quad \llcorner \; FIf := 0$

Algorithm 3: Create & optimize fuzzy rule for classification Algorithm

---

Input : Fuzzy Importance factor matrix and Linguistic gene expression matrix

Output: optimized fuzzy rule for classification of samples.

repeat

$\quad |\quad$ select a set of gene having maximum FIF;

$\quad |\quad$ create fuzzy rule according to gene set and corresponding its linguistic value;

$\quad |\quad$ Evaluate classification accuracy;

$\quad |\quad$ Raise new condition(gene)

$\quad |\quad$ Ultimate fuzzy rule having maximum classification accuracy used for sample

$\quad |\quad$ classification.

until *Classification accuracy not improve OR*

no new fuzzy rule available;

Table 1 Comparison between OFRG and CFS algorithms on lung cancer data

Samples	CFS classification accuracy in %	OFRG classification accuracy in %
100	82	94
110	76.36	93.64
120	80	94.17
130	71.54	93.85
140	72.14	94.29
150	78.67	94.67
160	68.12	94.37
170	75.29	94.71
180	73.89	94.44

A ten-fold cross-validation strategy is used by weka [29] classifier where BayesNet applied to measure the generalization accuracy of the Correlation feature selection (CFS) algorithm, shown in Table 1.

CFS [30] algorithm couples evaluation formula with an appropriate correlation measure and a heuristic search strategy. CFS was evaluated by experiments on artificial and natural data sets using three machine learning algorithms: a decision tree learner, an instance based learner, and Naive Bayes. Experiments on artificial data sets show that CFS quickly identifies and screens irrelevant, redundant and noisy features, and selects relevant features as long as their relevance does not strongly depend on other features. On natural domains, CFS typically eliminates well over half the features.

The OFRG algorithm extracts 4 (four) genes out of 22215 using GSE4115 [27, 28] data set showing more promising result than CFS algorithm (select 23 genes). Classification percentage comparison of CFS and OFRG algorithms are shown in Table 1 and Fig. 1, respectively. McNemar's test using Chi squared and corrected for discontinuity comparison are shown in Table 2.

4 Conclusions

Dimension reduction is one of the main issues in microarray data classification and appropriate gene selection demonstrates promising outcome that enhance knowledge discovery and model interpretation.

Since biologists are often interested in identifying a collection of genes involved in a biological function or a pathway rather than individual genes, there has been considerable interest in recent years to develop statistical methods for identifying significant set of genes. This paper exploits structural information and proposes a three stage strategy for selecting significant set of genes and classifying disease and normal experimental conditions. Proposed OFRG algorithm selects optimized fuzzy rule base using maximum fuzzy importance factor of genes for disease sample which helps to identify samples in proper classes with less number

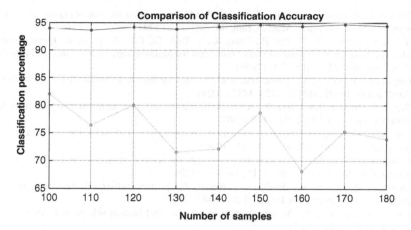

Fig. 1 Comparison of classification accuracy in percentage using OFRG and CFS algorithms on lung cancer data

Table 2 McNemar's test using Chi squared and P value of OFRG algorithm applied on lung cancer data

Total instances	OFRG Chi^2	OFRG P
100	0.1667	0.683091
110	0	1
120	0	1
130	0.1250	0.723674
140	0.4857	0.1250
150	0.1250	0.723674
160	0.4444	0.504985
170	0.4444	0.504985

of comparison. Moreover, variation of number of samples does not affect the classification accuracy of OFRG algorithm while it reduces in case of CFS algorithm.

References

1. Kononenko, I.: Inductive and bayesian learning in medical diagnosis. Appl. Artif. Intell. **7**(4), 317–337 (1993)
2. Wolberg, W., Street, W.: Mangasarian ol. machine learning techniques to diagnose breast cancer from fine-needle aspirates. Cancer Lett. **77**, 163–171 (1994)
3. Wolberg, W., Street, W.: Mangasarian ol. image analysis and machine learning applied to breast cancer diagnosis and prognosis. Anal. Quant. Cytol. Histol. **17**(2), 77–87 (1995)
4. Kurgan, L., Cios, K., Tadeusiewicz, R., Ogiela, M., Goodenday, L.S.: Knowledge discovery approach to automated cardiac spect diagnosis. Artif. Intell. Med. **23**(2), 149–169 (2001)
5. Antoniadis, A., Lambert-Lacroix, S., Leblanc, F.: Effective dimension reduction methods for tumor classification using gene expression data. Bioinformatics **19**(5), 563–570 (2003)

6. Guyon, I., Weston, J., Barnhill, S., Vapnik, V.: Gene selection for cancer classification using support vector machines. Mach. Learn. **46**, 389–422 (2002)

7. Yu, J., Ongarello, S., Fiedler, R., Chen, X., Toffolo, G., Cobelli, C., et al.: Ovarian cancer identification based on dimensionality reduction for high-throughput mass spectrometry data. Bioinformatics **21**, 2200–2209 (2005)

8. Oh, I., Lee, J., Moon, B.: Hybrid genetic algorithms for feature selection. IEEE Trans. Pattern Anal. Mach. Intell. **26**(11), 1424–1437 (2004)

9. Saeys, Y., Inza, I., Larranaga, P.: A review of feature selection techniques in bioinformatics. Bioinformatics **23**(19), 2507–2517 (2007)

10. Liu, H., Motoda, H.: Feature extraction, construction and selection: a data mining perspective, 1st edn. Kluwer, Norwell (1998)

11. Conilione, P., Wang, D.: A comparative study on feature selection for E. coli promoter recognition. Int. J. Inf. Technol. **11**, 54–66 (2005)

12. Degroeve, S., Baets, B., de Peer, Y., Rouzé, P.: Feature subset selection for splice site prediction. Bioinformatics **18**(Suppl 2), 75–83 (2002)

13. Guyon, I., Elisseeff, A.: An introduction to variable and feature selection. J. Mach. Learn. Res. **3**, 1157–1182 (2003)

14. Liu, H., Yu, L.: Toward integrated feature selection algorithms for classification and clustering. IEEE Trans. Knowl. Data Eng. **17**(4), 491–502 (2005)

15. Kuncheva, L.: Fuzzy Classifier Design. Springer, Heidelberg (2000)

16. Leondes, C. (ed.): Fuzzy Theory Systems: Techniques and Applications, vol. 1–4. Academic Press, San Diego (1999)

17. Yuan, Y., Shaw, M.: Induction of fuzzy decision trees. Fuzzy Sets Syst. **25**, 125–139 (1995)

18. Ichihashi, H., Shirai, T., Nagasaka, K., Miyoshi, T.: Neuro-fuzzy ID3: a method of inducing fuzzy decision trees with linear programming for maximizing entropy and an algebraic method for incremental learning. Fuzzy Sets Syst. **84**, 1–19 (1996)

19. Yuan, Y., Zhuang, H.: A genetic algorithm for generating fuzzy classification rules. Fuzzy Sets Syst. **84**, 1–19 (1996)

20. Castillo, L., Gonzalez, A., Perez, P.: Including a simplicity criterion in the selection of the best rule in a genetic fuzzy learning algorithm. Fuzzy Sets Syst. **120**(2), 309–321 (2001)

21. Castro, J., Castro-Schez, J., Zurita, J.: Use of a fuzzy machine learning technique in the knowledge acquisition process. Fuzzy Sets Syst. **123**(3), 307–320 (2001)

22. Jin, Y.: Fuzzy modeling of high-dimensional systems: complexity reduction and interpretability improvement. IEEE Trans. Fuzzy Syst. **8**(2), 212–221 (2000)

23. de Oliveira, V.: Semantic constraints for membership function optimization. IEEE Trans. Syst. Man Cybern. Part A Syst. Hum. **29**(1), 128–138 (1999)

24. Pedrycz, W., de Oliveira, V.: Optimization of fuzzy models. IEEE Trans. Systems Man Cybern. Part B Cybern. **26**(4), 627–637 (1996)

25. Setnes, M., Babuska, R., Verbruggen, B.: Rule-based modeling: precision and transparency. IEEE Trans. Systems Man and Cybern. Part C Appl. Rev. **28**(1), 165–169 (1998)

26. Setnes, M., Roubos, H.: GA-fuzzy based modeling and classification: complexity and performance. IEEE Trans. Fuzzy Syst. **8**(5), 509–522 (2000)

27. Spira, A., Beane, J., Shah, V., Steiling, K., Liu, G., Schembri, F., Gilman, S., Dumas, Y., Calner, P., Sebastiani, P., Sridhar, S., Beamis, J., Lamb, C., Anderson, T., Gerry, N., Keane, J., Lenburg, M., Brody, J.: Airway epithelial gene expression in the diagnostic evaluation of smokers with suspect lung cancer. Nat. Med. **13**(3), 361–366 (2007)

28. Gustafson, A., Soldi, R., Anderlind, C., Scholand, M., Qian, J., Zhang, X., Cooper, K., Walker, D., McWilliams, A., Liu, G., Szabo, E., Brody, J., Massion, P., Lenburg, M., Lam, S., Bild, A., Spira, A.: Airway PI3K pathway activation is an early and reversible event in lung cancer development. Sci. Transl. Med. **2**(26), 26–25 (2010)

29. Hall, M., Frank, E., Holmes, G., Pfahringer, B., Reutemann, P., Witten, I.: The weka data mining software: an update. SIGKDD Explor. **11**, 10–18 (2009)

30. Hall, M.: Correlation-based feature selection for machine learning. Thesis for the degree of Doctor of Philosophy (1999)

Comparative Study of Radial Basis Function Neural Network with Estimation of Eigenvalue in Image Using MATLAB

Abhisek Paul, Paritosh Bhattacharya and Santi Prasad Maity

Abstract Radial Basis Functions (RBFs) are very important in neural network. In this paper various Radial Basis Functions of neural network such as Generalized Inverse Multi Quadratic, Generalized Multi Quadratic and Gaussian are compared with matrix of images. Mathematical calculation, comparative study and simulation of Eigen value of matrix show that Gaussian RBF performs better result and gives lesser error compared to the other Radial Basis Functions of neutral network.

Keywords Gaussian · Generalized-multi-quadratic · Generalized-inverse-multi-quadratic · Neural network · Radial basis function

1 Introduction

In neural networks different Radial Basis Functions are utilized in various areas such as approximation process, pattern recognition, optimization etc. Out of many neural networks we have taken some Radial Basis Function neural network calculation. Eigen value is the characteristic root of any equation. In this paper we have introduced three RBF neural network methods to compare and calculate Eigen values of matrixes of images. Gaussian RBF, Generalized-Multi-Quadratic and Generalized-Inverse-Multi-Quadratic are introduce compute and compare with traditional methods. Some tuneable parameters are there for the generalized techniques such as for Generalized-Multi-Quadratic and α for Generalized-

A. Paul (✉) · P. Bhattacharya
Department of Computer Science and Engineering, National Institute of Technology, Agartala, India
e-mail: abhisekpaul13@gmail.com

S. P. Maity
Department of Information Technology, Bengal Engineering and Science University, Shibpur, India

G. P. Biswas and S. Mukhopadhyay (eds.), *Recent Advances in Information Technology*, 141
Advances in Intelligent Systems and Computing 266, DOI: 10.1007/978-81-322-1856-2_16,
© Springer India 2014

Fig. 1 Radial basis function
neural network architecture

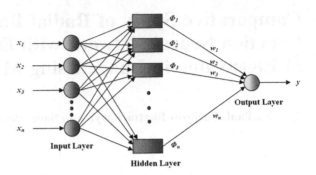

Inverse-Multi-Quadratic. In this paper matrix of some images and their YUV color space matrixes are given for computational process [1–3].

Section 2 describes of architecture of RBF neural network. In Sect. 3 experimental processes are given. Section 4 shows simulation and finally in Sect. 5 conclusion is given.

2 Architecture of Radial Basis Function Neural Network

RBF neural network architecture is of three layers, namely input, hidden and output. In Fig. 1 inputs are x_1, x_2, $x_3, \ldots,$ x_n which enter into input layer. Radial centres are $c_1, c_2, \ldots, c_n$. In hidden layer $\Phi_1, \Phi_2, \Phi_3, \ldots, \Phi n$ are the functions for distance between inputs and centres. Centres are of $n \times 1$ dimension when the number of input is n. The desired output is given by y which is calculated by proper selection of w_j. Here, w_j is the weight of jth centre. Summation of all $\Phi_i w_i$ is the output [1, 2] (Fig. 2).

$$y = \sum_{j=1}^{m} \phi_j w_j \tag{1}$$

$$\phi_j(x) = (||x - x_j||) \tag{2}$$

Radial basis functions such as Gaussian, Generalized-Multi-Quadratic and Generalized-Inverse-Multi-Quadratic are given below in Eqs. (3, 4, and 5). Euclidian distance and maximum distance from centres are z and σ.

- Gaussian:

$$\phi(z) = e^{-z^2/2\sigma^2} \tag{3}$$

- Generalized Multi-Quadratic:

$$\phi(z) = (z^2 + r^2)^\beta, 1 > \beta > 0 \tag{4}$$

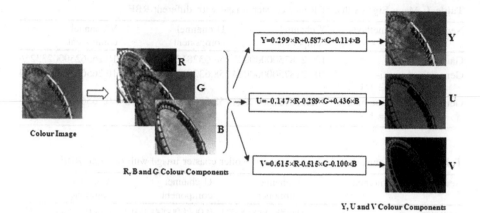

Fig. 2 Conversion of *Red*, *Green* and *Blue* components of colour image to Y, U and V components

- Generalized Inverse-Multi-Quadratic:

$$\phi(z) = (z^2 + r^2)^{-\alpha}, \alpha > 0 \tag{5}$$

3 Experimental Analysis

First of all we have taken colour images. Then we extracted red, green and blue channel components matrix from the colour image. We then calculated Y, U and V colour space component [4] matrix from the red, green and blue component matrix of the respective image. In Eqs. (6, 7 and 8) relation between of red, green blue channel component and Y, U and V channel component is given. As we have applied Gaussian, Generalized-Multi-Quadratic and Generalized-Inverse-Multi-Quadratic RBF neural network methods. RBF need optimal selections weight and the centres parameters. Here, we applied pseudo inverse technique [5] to process the parameters. Size of matrix is 128×128 pixels each for every experiment.

$$Y = 0.299\,R + 0.587\,G + 0.114\,B \tag{6}$$

$$U = -0.147\,R - 0.289\,G + 0.436\,B \tag{7}$$

$$V = 0.615\,R - 0.515\,G - 0.100\,B \tag{8}$$

In normal method Eigen values of matrix in Y, U and V channel component image are 1.002187500000001, 58.632812499999964 and 10.265624999999984 respectively.

Table 1 Mean Eigen value of Roller coaster image with different RBF

Neural network methods	Y channel component	U channel component	V channel component
Gaussian	100.2187500000022	58.632812500000092	10.265625000233256
Generalized-multi-quadratic, $\beta = 1/4$	100.2187500000024	58.632812499997314	10.265625015438804
Generalized inverse-multi-quadratic, $\alpha = 1/4$	100.2187500000044	58.632812499999162	10.265624973651665

Table 2 Error comparison in Eigen value of Roller coaster image with different RBF

Neural network methods	Y channel component	U channel component	V channel component
Gaussian	0.0000000000021	0.000000000000128	0.000000000233272
Generalized-multi-quadratic, $\beta = 1/4$	0.0000000000023	0.000000000000265	0.000000001543882
Generalized inverse-multi-quadratic, $\alpha = 1/4$	0.0000000000043	0.000000000000802	0.000000026348319

In normal method we have calculated Eigen values of Y, U and V channels component images matrix of rose image are 138.5625000000003, 44.539062500000071and 56.859374999999915 respectively (Tables 1 and 2).

4 Simulation

Two different colour images namely roller coaster [6] and rose [7] and their YUV colour space images are taken as example for experiment. We have simulated Eigen value of matrixes of these images with normal method, Gaussian method, Generalized-Inverse-Multi-Quadratic method and Generalized-Multi-Quadratic method. We have used MATLAB 7.6.0 software [8] for the simulation process. In Fig. 3 Roller coaster image and its corresponding YUV channel component's images are shown. In Fig. 4 Rose image and its corresponding YUV channel component's images are given. For Fig. 3 values of β and α are 1/4. For Fig. 4 values of β and α are 1/5. Average or mean Eigen values of matrixes of all these images with Gaussian RBF, Generalized-Inverse-Multi-Quadratic RBF and Generalized-Multi-Quadratic RBF are calculated and simulated. These values are shown in Tables 1 and 3. Corresponding relative errors of Eigen values of matrixes are given in Tables 2 and 4.

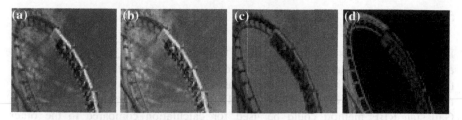

Fig. 3 Roller coaster image **a** Original color Image, **b** Y channel component, **c** U channel. component, **d** V channel component

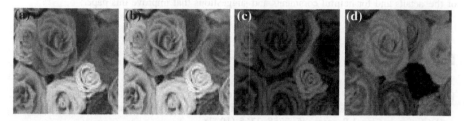

Fig. 4 Rose image **a** Original color Image, **b** Y channel component, **c** U channel component, **d** V channel component

Table 3 Mean Eigen value of Roller coaster image with different RBF

Neural network methods	Y channel component	U channel component	V channel component
Gaussian	138.5625000000003	44.539062500000071	56.859374999999872
Generalized-multi-quadratic, $\beta = 1/5$	138.5625000000016	44.539062500000092	56.859374999999169
Generalized inverse-multi-quadratic, $\alpha = 1/5$	138.5625000000001	44.539062499999801	56.859375000000732

Table 4 Error comparison in Eigen value of Roller coaster image with different RBF

Neural network methods	Y channel component	U channel component	V channel component
Gaussian	0.0000000000000	0.000000000000000	0.000000000233272
Generalized-multi-quadratic, $\beta = 1/4$	0.0000000000020	0.000000000000021	0.000000001543882
Generalized inverse-multi-quadratic, $\alpha = 1/4$	0.0000000000002	0.000000000000270	0.000000026348319

5 Conclusion

In this paper various RBF methods like Generalized-Multi-Quadratic RBF Generalized-Inverse-Multi-Quadratic RBF and Gaussian RBF methods are utilized for the computation of Eigen values of various matrixes. Simulation results give better result and lesser error for Gaussian RBF method. So, it can be believed that Gaussian RBF method could be used for calculation compared to the other methods in neural network.

Acknowledgments The authors are so grateful to the anonymous referee for a careful checking of the details and for helpful comments and suggestions that improve this paper.

References

1. Schölkopf, B., Sung, K.-K., Burges, C.J.C., Girosi, F., Niyogi, P., Poggio, T., Vapnik, V.: Comparing support vector machines with Gaussian kernels to radial basis function classifiers. IEEE Trans. Signal Process. **45,** 2758–2765 (1997)
2. Mao, K.Z., Huang, G.-B.: Neuron selection for RBF neural network classifier based on data structure preserving criterion. IEEE Trans. Neural Networks **16**(6), 1531–1540 (2005)
3. Luo, F.L., Li, Y.D.: Real-time computation of the eigenvector corresponding to the smallest eigen value of a positive definite matrix. IEEE Trans. Circuits Syst. **1**(41), 550–553 (1994)
4. Noda, H., Niimi, M.: Colorization in YCbCr color space and its application to JPEG images. Pattern Recogn. **40**(12), 3714–3720 (2007)
5. Klein, C.A., Huang, C.H.: Review of pseudo-inverse control for use with kinematically redundant manipulators. IEEE Trans. Syst. Man Cybern. **13**(3), 245–250 (1983)
6. http://www.nynyhotelcasino.com/images/attractions/the-roller-coaster-large.jpg
7. http://ibmsmartercommerce.sourceforge.net/wp-content/uploads/2012/09/Roses_Bunch_Of_Flowers.jpeg
8. Math works.: MATLAB 7.6.0 (R2008a) (2008)

Author Index

B
Babu, Ramesh, 95
Bandyopadhyay, Arnab, 109
Bera, Sahadev, 37
Bhattacharya, Bhargab B., 37
Bhattacharya, Paritosh, 141
Bhunia, Suman Sankar, 19
Biswas, Arindam, 37
Biswas, G. P., 3, 11, 27
Bose, Mahua, 125

C
Choudhury, Suvra jyoti, 57

D
das Mukhopadhyay, Chitrangada, 133
Das, Asit Kumar, 57
Das, Sunanda, 57
Datta, Amit, 87
De, Mallika, 87
Dhar, Sourav Kumar, 19

H
Hangarge, Mallikarjun, 49

I
Islam, S. K. Hafizul, 27

K
Kare, Anjeneya Swami, 77

M
Maity, Santi Prasad, 141
Malwe, Shweta R., 3
Mishra, Manoj K., 27
Mukherjee, Nandini, 19
Mukherjee, Sankar, 11

N
Naik, Kshiramani, 65

P
Pal, Arup Kumar, 65
Paul, Abhisek, 141
Paul, Amit, 133

R
Rao, Subba, 95
Roy, Parimita, 117
Roy, Sarbani, 19

S
Santosh, K. C., 49
Saxena, Sanjeev, 77
Sharma, Arpita, 125
Sil, Jaya, 57, 133
Singh, Garib Nath, 109

U
Upadhyay, Ranjit Kumar, 117

G. P. Biswas and S. Mukhopadhyay (eds.), *Recent Advances in Information Technology*,
Advances in Intelligent Systems and Computing 266, DOI: 10.1007/978-81-322-1856-2,
© Springer India 2014